Powering Humanity

Essays on Energy and Society

Michael E. Webber

Author of *Thirst for Power* and *Power Trip*

Cover art by Jeffrey M. Phillips. Interior art by David D. Webber.

ISBN: 979-8-35093-387-1 (print)
ISBN: 979-8-35093-388-8 (eBook)

Contents

WESTMINSTER ABBEY,
LONDON

Acknowledgments

THIS COLLECTION OF ESSAYS UNFOLDED PIECE-BY-PIECE OVER fourteen years and there are many people who helped make them a reality and deserve acknowledgment. To start with, thanks to the incredible researchers on my team at the University of Texas at Austin (UT) who performed countless analyses that revealed new facts and findings underpinning many of these pieces. Some of them – Kelly T. Sanders, Sheril R. Kirshenbaum, Alex C. Breckel, John R. Fyffe, Joshua D. Rhodes, Thomas A. Deetjen, F. Todd Davidson, Yael R. Glazer and Jamie J. Lee – were co-authors of some of the original versions of these essays. And many others, including Ashlynn S. Stillwell, Amanda Cuellar and Carey W. King, performed some of the foundational research. I have unending gratitude to Sarah De Berry-Caperton for making our UT team successful and productive. We couldn't do it without her.

My research sponsors over the years have also helped the team do nontraditional work that stitches together different elements of society. Some of these sponsors year in and year out include the Cynthia and George Mitchell Foundation, the Alfred P. Sloan Foundation, Itron, ENGIE and the Texas State Energy Conservation Office. Marilu Hastings deserves special acknowledgment for her vision and support of some of our most impactful work.

I would also like to thank The Rockefeller Foundation and the Bellagio Center, whose support of my public education efforts to improve energy and environmental literacy has been critical. Their sponsorship of the writer's residency program in Bellagio, Italy, made this collection a reality. Pulling together this compendium in such a beautiful spot surrounded by world-class intellectuals was a humbling and inspiring experience. Special thanks to Ashvin Dayal, Zia Khan, Bethany Martin-Breen, Sarah Geisenheimer, Monique Hoeflinger, Alice Luperto and Pilar Palacia, whose tireless efforts made for a productive and enjoyable creative experience.

Every writer needs an editor and I'm no exception. I want to thank the editors who have done the most to improve my writing of the essays: Megan Sever, Jeffrey Winters, JB Bird, Chris Conway and Mark Fischetti. Others who have edited my writing on multiple occasions include Alka Tripathy-Lang, Matt Pene and Jennifer Bogo. I am grateful to all of them, but Megan is especially deserving of my thanks for her heavy lifting for some of the original essays in *Geotimes* and *EARTH Magazine* and for this collection.

Many thanks also to Julia Webber, David E. Webber, David D. Webber, Stephanie Webber Perry, Maddie Taylor, Sheril Kirshenbaum, Linda Kirshenbaum, David Lowry, Atlas Lowry, Apollo Lowry, Rebecca Sewell, Katherine Spiering, Susie Temple, Alicia Groos, Cuatro Groos, Dennis McWilliams, Jennifer Bristol, Andrew Brown, Marilu Hastings, Megan Sever, and Alka Tripathy-Lang for helping me figure out the book's title, scope or intent, and motivating me to put words to paper.

Thanks to Jeff Phillips for the cover art. His designs always hit the mark.

Many thanks to David D. Webber for creating the art for the chapter headings.

I've already thanked her, but Julia Webber deserves additional recognition (and love!) for all her patience and encouragement over the last three decades while I tackled the original research projects, wrote hundreds of essays, and worked on pulling them together for this publication. She is the motivation and inspiration for much of my effort and I appreciate her kindness and steadfast support over all these years.

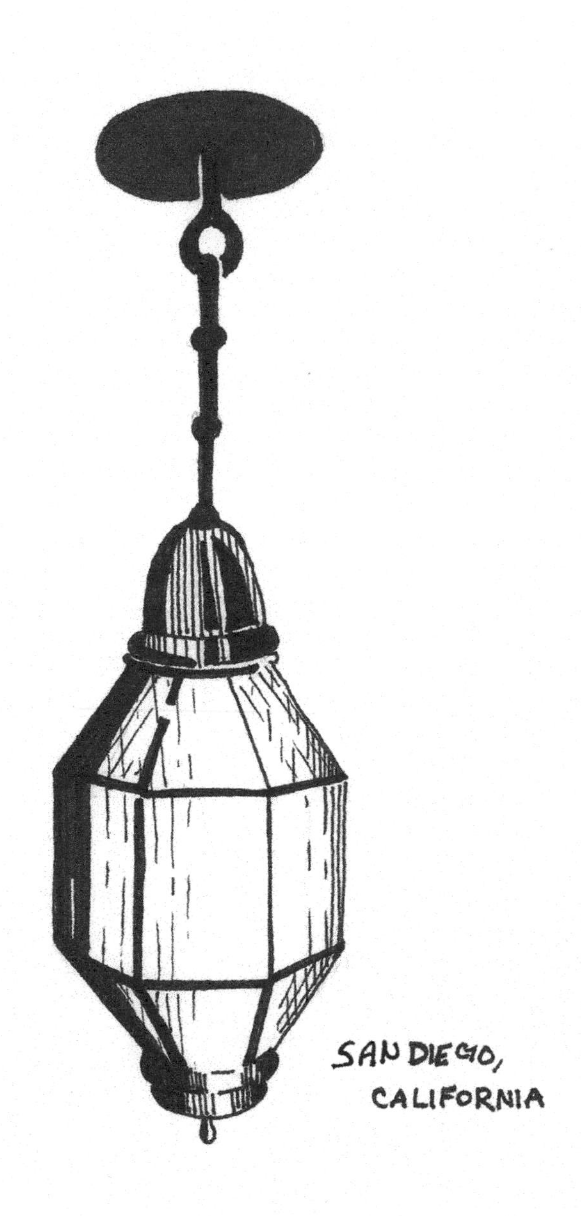
SAN DIEGO,
CALIFORNIA

Prologue: Energy Interconnections

AS CLASSICALLY DEFINED FROM THE EARLY INDUSTRIAL ERA, energy is "the capacity to do work." However, in the modern context, that definition seems very limited compared with what energy actually offers society. Taking a broader and updated view, energy could be defined as "the ability to do interesting and useful things." Energy brings illumination, information, heat, clean water, abundant food, motion, comfort and much more to our homes and factories with the turn of a valve or the touch of a button. It is the potential to harvest a crop, refrigerate it, and fly it around the world. It is the ability to drive across the country or fly across the world in the fraction of the time it would take to walk, ride or row. And it also facilitates education, health and security.

Energy's importance was noted by late Nobel laureate Richard E. Smalley of Rice University in a lecture he gave in 2003 highlighting the "Top Ten Problems of Humanity for the Next 50 Years." His list was organized in descending order of importance, with energy at the top. Developing plentiful sources of clean, reliable, affordable energy, he argued, enables us to tackle all the subsequent problems of humanity, related to water, food, democracy, war, and so forth. I agree. Energy is a vital part of our world.

Our civilization is founded on access to energy; the corollary is therefore that a lack of energy would lead to its collapse.

Energy cuts across all the sectors – water, food, health, security, the environment and the economy – that matter for a peaceful, prosperous life. It is complex, intertwined and dynamic. It is connected to all parts of our lives and societies and, arguably, we couldn't live without it. In that way, energy itself has become one of our most basic needs. Energy is also part of every one of the five main human needs.

For that reason, I have organized this compendium of essays I've written over the past decade and a half according to Maslow's hierarchy of needs:

> Physiological needs: Energy, food and water;
> Garbage in and garbage out
>
> Safety needs: Energy and education;
> Energy and the environment
>
> Connection needs: Energy and the global economy;
> Energy and transportation
>
> Esteem needs: Old versus new energy; Energy and innovation.
>
> Self-Actualization needs: Energy and society;
> Energy and culture; The folly of predictions.

For developing economies, energy is often still a wish. Energy, water, food, sanitation and waste disposal – they're not givens. For modern economies with universal energy access, it is easy to take for granted that energy will always be there – that our physiological needs will be met. More than 99 percent of the time in the United States, we don't think about these services: The water flows at the turn of a valve, the wastewater tidily drains away at the push of a button, the lights shine at the flick of a switch, and the gas warms our homes, heats our water or cooks our food with the twist of a knob. This is the way it should be.

When storms or other natural disasters strike though, and the power, water, gas or gasoline gets disrupted, then the value of ready access to energy becomes clear. This sentiment is captured by the adage that "we know the value of water when the well is dry." Likewise, we know the value of energy when our lights are out or we can't get fuel at the pump.

At major holidays, the water and wastewater treatment plants don't shut down. When pandemics sweep across the world, when natural disasters hit or economic markets stumble, broken water and sewer pipes don't wait for repair. Power outages don't wait to be fixed next week. Usually, someone is away from their family taking care of those problems for us. Like doctors and nurses who work holidays, so, too, do energy workers. I would argue that the biggest threat to public health has everything to do with whether the water and wastewater systems are operating properly. If water systems fail, public health crises can escalate out of control quickly–as we see occurs frequently in developing economies in the wake of natural disasters, like when a cholera outbreak struck Haiti after the devastating 2010 earthquake. Or when the power went out for several days across Texas in 2021 due to a cold snap; that disaster ended up killing at least 246 people, but possibly several hundred more.

So when we think about thanking our frontline heroes, like doctors and nurses, we should also think about our energy workers, arguably, modern society's most important people – invisible superheroes.[1] But instead of capes and masks, they wear hardhats and steel-toed shoes.

Throughout the COVID-19 pandemic, natural disasters and severe weather, energy workers have been hard at work in dangerous conditions, clearing debris, stringing lines and putting themselves at risk to make our lives more comfortable. This heroism is highlighted in the 2022 Russian invasion of Ukraine. Russia's attacks on energy infrastructure put Ukrainian people at risk beyond just the collateral damage of the bombs, bullets and missiles. Energy workers worked to bring systems back online, putting their own lives at risk.

Utility workers are invisible when they are doing their jobs well. In a modern society, the best utilities are the ones we don't think about. If we think about a utility, it's usually for negative reasons: The bills are

---

1 Portions of this prologue are adapted from "Invisible Heroes," by Michael E. Webber, *Mechanical Engineering*, May 2020.

too high; the service is unreliable; or its system sparked fires, spills, contaminations or explosions.

It is from that lens – the view that energy is essential – that I authored or co-authored some 200 essays, op-eds and contributions for newspapers, magazines and blogs from 2008 to 2022. There are some recurring themes: energy's interconnections with everything we care about, a need for rational policies, geopolitical impacts from local events and persistent change. There are some predictions. Some of these essays in retrospect look prescient while others look overly simplistic or short-sighted. But even the ones that were flat-out wrong are useful because they reveal what I was thinking (or what conventional wisdom held to be true) at the time, even if the future eventually proved us wrong. The mistakes offer useful lessons about how we should not underestimate the potential for change, innovation and unforeseen events.

As someone engaged with the general public, policymakers and industry, I felt compelled to write those articles over the years to clarify for myself my own thoughts, but also to inject some more viewpoints into the public discourse. Hopefully those contributions were helpful.

Organizing such a vast body of words is no easy task. Fewer than half of the essays I wrote between 2008 and 2022 are included in this collection, and we updated numbers in those that are included.

Energy is a dynamic industry, so important events and trends unfolded during the time span that these essays cover. Oil prices spiked, collapsed, rose, collapsed again, then spiked again. Wind, solar and battery prices plunged as market adoptions soared. Disasters such as the 2011 earthquake and tsunami in Japan that triggered a failure of the nuclear power plant in Fukushima, the 2010 Deepwater Horizon blowout in the Gulf of Mexico, the 2017 near-complete dam collapse in Oroville, California, plus extreme weather events like droughts and hurricanes, significant geopolitical events such as the COVID-19 pandemic, civil unrest, refugee crises, and land wars all strained and pushed the energy system in ways that were hard to anticipate. The consequences of those events continue to cascade globally.

Market forces and policies are additional external factors that impose change on the energy industry, as consumers, investors and policymakers demand more from the system, pushing for greater cleanliness, reliability, affordability, and in some cases domestic sourcing.

Another major theme my essays dive into is energy transitions. Change is in fact one of the constants of the energy industry – it is always changing. New technologies improve the way we produce and move energy. New appliances change where and why we use energy. Evolving concerns about energy reliability, geopolitical risks and environmental legacies shift our priorities over time to cleaner and more secure forms of energy. Taken together, this collection of changes gives us an "energy transition." The energy world has always been one of transitions: In the second half of the 1800s we moved from wood to coal and whale oil to kerosene for heat and lighting. For power generation, we started with falling water then added steam power (from coal, nuclear, gas, oil, geothermal or wood sources) before adding wind and solar power over the last century-plus. Now, we're in the process of transitioning away from unscrubbed carbon-based energy that produces greenhouse gases we dump in the sky to cleaner forms and processes as we face the climate crisis that might displace hundreds of millions of people while affecting ecosystems, coastlines, aquifers and agriculture. Though climate change is driven by many factors including land use changes and agriculture, the way we produce, move and consume energy is responsible for about two-thirds of the overall effect.

Unfortunately, the energy industry has a reputation for being slow moving. So a question on the table is how this transition – which is a combination of changing forms of energy (for example from fossil fuels to other options) and changing technologies (such as the shift from combustion engines to electric vehicles) – can be accelerated to meet our climate change targets without compromising quality of life.

The slow pace of regulators and major energy companies doesn't have to hold us back. The energy space is evolving with smarter platforms and more technology. It is where new solutions that leverage extensive data and computing tools such as artificial intelligence are finding applications. It's also a field relying on ubiquitous sensors and innovations to lower cost, reduce emissions, and improve safety and reliability – even in the face of more severe climate crises. That means our heroes will need more specialized training.

When younger workers say they want to work in clean tech, energy is where the action is. We should all encourage our students, supervisees, children and even colleagues to take another look at the energy sector as a place of innovation and fulfillment. It is where solutions can take root to improve the plight of humanity. If our best and brightest don't go into the energy sector, then we will eventually pay the price.

Many of these essays were written with those people in mind.

The essays in this compendium are not in chronological order. You should not feel it necessary to read this book in sequence or cover-to-cover but should feel free to skip around and read whatever you find most interesting. Unless I list a co-author, each of the pieces in this collection was written only by me.

Enjoy these musings and may your energy systems – wherever you are – be clean, cheap and reliable.

Section I

Physiological Needs

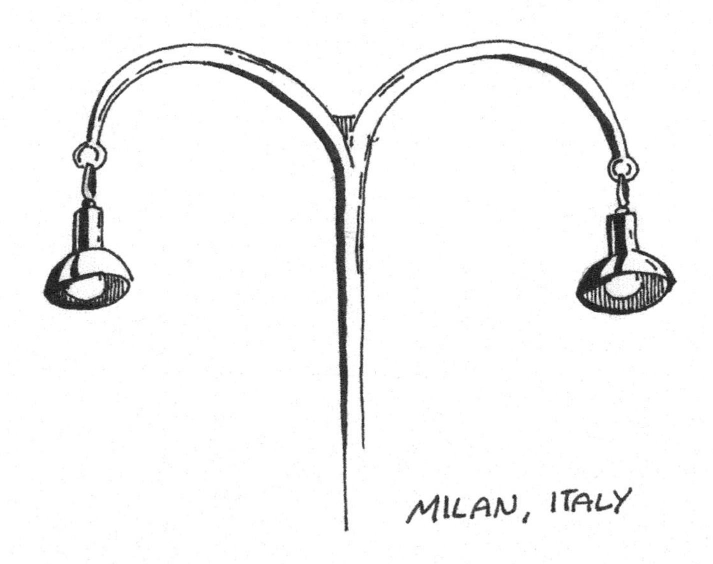
MILAN, ITALY

Chapter 1

Physiological Needs I: Energy, Food & Water

Our Future Rides on Our Ability to Integrate Energy + Water + Food: A Puzzle for the Planet

SCIENTIFIC AMERICAN, FEBRUARY 2015

> *The most important innovation we need is holistic thinking about all our resources.*

IN JULY 2012, THREE OF INDIA'S REGIONAL ELECTRIC GRIDS FAILED, triggering the largest blackout on Earth. More than 620 million people – 9 percent of the world's population – were left powerless. The cause: the strain of food production from a lack of water. Because of major drought, farmers plugged in more and more electric pumps to draw water from deeper and deeper below ground for irrigation. Those pumps, working furiously under the hot sun, increased the demand on power plants. At the same time, low water levels meant hydroelectric dams were generating less electricity than normal.

Making matters worse, runoff from those irrigated farms during floods earlier in the year left piles of silt right behind the dams, reducing the water capacity in the dam reservoirs. Suddenly, a population larger than all of Europe and twice as large as that of the U.S. was plunged into darkness.

California is facing a surprisingly similar confluence of energy, water and food troubles. Reduced snowpack, record-low rainfall and ongoing development in the Colorado River basin have reduced the river water in Central California by a third. The state produces half of the country's fruits, nuts and vegetables and almost a quarter of its milk, and farmers are pumping groundwater like mad; last summer some areas pumped twice as much water for irrigation as they did the previous year. The 400-mile-long Central Valley is literally sinking as groundwater is pulled up from below. Just when more power is needed, Southern California Edison shut down two big nuclear reactors for a lack of cooling water. San Diego's plan to build a desalination plant along the coast was challenged by activists who opposed the facility on the grounds that it would consume too much energy.

Energy, water and food are the world's three most critical resources. Although this fact is widely acknowledged in policy circles, the interdependence of these resources on one another is significantly underappreciated. Strains on any one can cripple the others. This situation has made our society more fragile than we imagine, and we are not prepared for the potential disaster that is waiting for us.

Yet we are making once-in-a-generation decisions about power plants, water infrastructure and farmland that will last for many decades, locking us into a vulnerable system. Meeting the world's energy needs alone will require $48 trillion in investment between 2015 and 2035, according to a 2014 International Energy Agency report, and the agency's executive director said there is a real risk "that investments are misdirected" because impacts are not being properly assessed.[2]

An integrated approach to solving these enormous issues is urgently needed rather than an attempt to solve each problem apart from the others. A vast number of the planet's population centers are hit with drought; energy systems are bumping up against environmental constraints and rising costs; and the food system is struggling to keep up with rapidly growing demand. And the nexus of food, water and energy is a backdrop to much of the most troubled parts of the world. Riots and revolutions in Libya and Syria were provoked by drought or high food prices, toppling governments. We need to solve the interconnected conundrum to create a more integrated and resilient society, but where do we start?

Cascading Risks or Rewards

The late Nobel laureate Richard E. Smalley of Rice University gave a hint at where to begin in his 2003 lecture highlighting the "Top Ten Problems of Humanity for the Next 50 Years." His list was organized in descending order of importance: energy, water, food, environment, poverty, terrorism and war, disease, education, democracy and population. Energy, water and food were at the top because solving them would combat problems lower down, in cascading fashion. Developing plentiful sources of clean, reliable, affordable energy, for example, enables an abundance of clean water. Having an abundance of clean water and energy (to make fertilizer and to power tractors) enables food production. And so on.

As brilliant as Smalley's list was, it missed two important nuances. First, energy, water and food are interconnected. And second, although

2 In 2021, the International Energy Agency's World Energy Outlook included an update, estimating that energy transition-related investments would have to increase to approximately $4 trillion per year by 2030 and beyond to get the world on track for reaching net-zero by 2050. Source: www.iea.org/weo

an abundance of one enables an abundance of the others, a shortage of one can create a shortage of the others.

With infinite energy, we have all the water we need because we can desalt the oceans, drill very deep wells and move water across continents. With infinite water, we have all the energy we need because we can build widespread hydroelectric plants or irrigate unlimited energy crops. With infinite energy and water, we can make the deserts bloom and build highly productive indoor farms that produce food year-round.

We do not live in a world with infinite resources, of course. We live in a world of constraints. The likelihood that these constraints will lead to cascading failures grows as pressure rises from population growth, longer lifespans and increasing consumption.

For example, Lake Mead outside Las Vegas, fed by the Colorado River, is now at its lowest level in history.[3] The city draws drinking water from what amounts to two big straws that dip into the lake. If the level keeps dropping, it may sink lower than those straws: large farming communities downstream could be left dry, and the huge hydroelectric turbines inside the Hoover Dam on the lake would provide less power or might stop altogether. Las Vegas's solution is to spend nearly $1 billion on a third straw that will come up into the lake from underneath. It might not do much good. Scientists at the Scripps Institution of Oceanography in La Jolla, Calif., have found that Lake Mead could dry up by 2021[4] if the climate changes as expected and cities and farms that depend on the Colorado River do not curtail their withdrawals.

---

3 Since this article was written in 2015, Lake Mead's level dropped further. According to NASA's Earth Observatory, in 2022, Lake Mead dropped to its lowest level since the reservoir was first built. Source: https://earthobservatory.nasa.gov/images/150111/lake-mead-keeps-dropping [Accessed November 12, 2022]

4 Though the scientists' analysis seemed quite stark, Lake Mead did not go dry in 2021. However, it got quite close, dropping to about 25% capacity in 2022 according to the NASA Earth Observatory article in the preceding footnote.

In Uruguay, politicians must confront tough decisions about how to use the water in their reservoirs. In 2008, the Uruguay River behind the Salto Grande Dam dropped to very low levels. The dam has almost the same electricity-generating capacity as the Hoover Dam, but only three of the 14 turbines were spinning because local people wanted to store the water for farming or municipal use. The citizens along the river and their political leaders were forced to choose whether they wanted electricity, food or drinking water. Constraints in one sector triggered constraints in the others. Although that threat might have temporarily eased for Uruguay, it repeats itself in other parts of the world. In like manner, certain communities in drought-stricken Texas and New Mexico have prohibited or restricted water for use in fracking for oil and gas, saving it for farming.

About 80 percent of the water we consume in the U.S. is for agriculture – our food. Nearly 13 percent of energy production is used to fetch, clean, deliver, heat, chill and dispose of our water. Fertilizers made from natural gas, pesticides made from petroleum, and diesel fuel to run tractors and harvesters drive up the amount of energy it takes to produce food. Food factories requiring power-hungry refrigeration produce goods wrapped in plastic made from petrochemicals, and it takes still more energy to get groceries from the store and cook them at home. The nexus is a big mess, and the entire system is vulnerable to a perturbation in any part.

Technical Solutions

It would be folly to build more power plants and water delivery and treatment facilities with the same old designs, to grow crops using the same outdated methods, and to extract more oil and gas without realizing that these pursuits impinge on one another. Thankfully, it is possible to integrate all three activities in ways that are sustainable.

The most obvious measure is to reduce waste. In the U.S., 25 percent or more of our edible food goes into the dump. Because we pour so much energy and water into producing food, reducing the proportion of waste can spare several resources at once. That might mean something as simple as serving smaller portions and eating less meat, which is much more energy intensive than grains. We can also put discarded food and agricultural waste such as manure into anaerobic digesters that turn it into biogas. These metal spheres look like shiny bubbles. Microbes inside break down the organic matter, producing methane in the process. If we implement this technology widely – at homes, grocery stores and central locations such as farms – that would create new energy and revenue streams while reducing the energy and water that are needed to process the refuse.

Wastewater is another byproduct we could turn into a resource. In California, San Diego and Santa Clara are using treated wastewater to irrigate land. Treating the water again to make it clean enough to drink could bolster municipal water supplies if regulators would allow it.

Urban farm proponents such as Dickson Despommier of Columbia University have designed "vertical farms" that would be housed inside glass skyscrapers. People in New York City, for example, produce a billion gallons of wastewater a day, and the city spends enormous sums to clean it enough to dump into the Hudson River. This cleansed water could instead irrigate crops inside a vertical farm, generating food while reducing the farm's demand for freshwater. Solids extracted from liquid waste are typically burned, but instead they could be incinerated to produce electricity for the big building, reducing its energy demand. And because fresh food would be grown right where many consumers live and work, less transportation would be needed to truck food in, potentially saving energy and carbon dioxide emissions.

Startup companies are trying to use wastewater and CO_2 from power plants to grow algae right next door. The algae eat the gas and water, and workers harvest the plants for animal feed and biofuel, all while tackling the fourth priority on Smalley's list – improving the environment – by removing compounds from the water and CO_2 from the atmosphere.

We could harness the same CO_2 to create energy. My colleagues at the University of Texas at Austin have designed a system in which waste CO_2 from power plants is injected into large brine deposits deep belowground. The CO_2 stays submerged, eliminating it from the atmosphere, and pushes out hot methane, which comes to the surface, where it can be sold for energy. The heat can also be tapped by industry.

Smart conservation is another way to spare different resources simultaneously. We use more water through our light switches and electrical outlets than our faucets and showerheads because so much water is needed to cool power plants that are out of sight and out of mind. We also use more energy to heat, treat and pump our water than we use for lighting. Turning off the lights and appliances saves vast amounts of water and turning off the water saves large amounts of energy.

We can also rethink how to better use energy and water to grow food in unlikely places. In parts of the Desert Southwest, brackish groundwater is abundant at shallow depths. Wind and solar energy are also plentiful. These energy sources present challenges to utilities because the sun does not shine at night and the wind blows intermittently. But that schedule is fine for desalting water because clean water is easy to store for use later. Desalination of seawater is energy intensive, but brackish groundwater is not nearly as salty. Our research at UT-Austin indicates that intermittent wind power is more economically valuable when it is used to make clean water from brackish groundwater than when it is used to make off-peak electricity. And of course, the treated water can then irrigate crops. This is the nexus working in our favor.

The same thinking can improve hydraulic fracturing for oil and gas. One unfortunate side effect is that waste gas, mostly methane, coming up the well is flared – burned off into the air. The flaring is so voluminous that it can be seen at night from space. The wells also produce a lot of dirty water – millions of gallons of freshwater injected into wells for fracking come back out laden with salts and chemicals. Operators can use the methane to power distillers or other heat-based machines to clean the water, making it reusable on-site, which spares freshwater while avoiding the wasted energy and emissions of a flare.

We can also be smarter about how we deliver water to homes and businesses. Sensors embedded in smart grids help to make electricity distribution more efficient. But our water system is a lot dumber than our electricity system. Outdated, century-old meters often fail to accurately record water use, and experts say that antiquated pipes leak 10 to 40 percent of the treated water that flows through them. Embedding wireless data sensors in the water delivery system would give utilities more tools to reduce the leaks – and lost revenues. Smart water would also help consumers manage their consumption.

We can do smart food too. One reason so much food is wasted is because grocery stores, restaurants and consumers rely on expiration dates, a crude estimate of whether food has spoiled. Food is not sold or consumed past the expiration date even though it may still be fine if its temperature and condition have been well managed. Using sensors to assess food directly would be smarter. For example, we could use special inks on food packaging that change color if they are exposed to the wrong temperature or if undesirable microbes begin to grow in the food, indicating spoilage. We can install sensors along the supply chain to measure trace gases that are released by rotting fruits and vegetables. Those same sensors can lead to tighter refrigeration controls that minimize losses.

New Policy Thinking

Although many technical solutions can improve the energy-water-food nexus, we often do not exploit them because ideologically and politically, the U.S. has not fully grasped the interrelatedness of these resources. Policymakers, business owners and engineers typically work in isolated fashion on one issue or another.

Sadly, we compound the problems with policy, oversight and funding decisions made by separate agencies. Energy planners assume they will have the water they need. Water planners assume they will have the energy they need. Food planners recognize the risks of drought, but their reaction is to pump harder and drill deeper for water. The most important innovation we need is holistic thinking about all our resources.

That kind of thinking can lead to smarter policy decisions. For example, policies can fund research into energy technologies that are water-lean, water technologies that are energy-lean, and food production, storage and monitoring techniques that prevent losses while reducing energy and water demands. Setting cross-resource efficiency standards can kill two birds with one stone. Building codes can also be a powerful tool for reducing waste and improving performance. Permitting for new energy sites should require water-footprint assessments, and vice versa. And policymakers can set up revolving loan funds, direct capital investments or tax benefits for institutions that integrate these kinds of technical solutions.

As Smalley pointed out, energy can be the driver. We must think about using our energy sector to solve multiple challenges simultaneously. Policies that are monomaniacal about lowering atmospheric CO_2 levels, for example, might push us toward low-carbon energy choices that are very water intensive, such as irrigated biofuels or coal plants with carbon capture.

Personal responsibility plays a role too. Demand for fresh salads that land on our winter plates from 5,000 miles away creates a far-flung, energy-hungry food distribution system. In general, our personal choices for more of everything just push our resources to the edge.

The energy-water-food nexus is the most vexing problem to face our planet. To quote the late George Mitchell, father of modern hydraulic fracturing and a sustainability advocate: “If we can’t solve the problem for seven billion people, how will we do so for nine billion people?”

The Energy-Water Nexus: Managing Water in an Energy Constrained World

EARTH MAGAZINE, JULY 2013

By Kelly T. Sanders and Michael E. Webber

Getting water to the right place at the right time with the right quality and temperature requires energy, and consequently, energy has become a constraining factor on our management of water issues.

WATER CAN BE TRICKY. WITH TOO LITTLE, CROPS DIE, INDUSTRIES move away, power plants fail, ecosystems suffer and people go thirsty. With too much, floods ruin infrastructure, destroy crops, spread water-borne diseases, and disrupt flows of clean water, wastewater, power and transportation. We want water at the right time and in the right place because moving and storing water require effort. We also want it at the right quality and the right temperature. Saline water, brackish water and polluted water are abundant but costly to treat. Water that's too warm won't cool power plants effectively and can damage ecosystems, while water that's too cold can burst pipes and damage infrastructure.

Water's capriciousness, however, can be tempered – with energy. If we had unlimited and perfectly clean energy (to mitigate environmental impacts) at our disposal, we could desalinate the ocean, providing enough potable water for everyone, everywhere. We could build pipelines to move water from where it is abundant to where it is scarce. We could build adequate sanitation infrastructure in communities that lack it so that raw sewage does not flow into freshwater ecosystems and cause sickness and eutrophication. We could build more storage reservoirs so that water could be collected in times of excess to be used in times of shortage.

Of course, we do not have unlimited energy and what we do have isn't perfectly clean, and consequently, energy has become a constraining factor on our management of water issues.

The corollary is also true: Just as energy constraints become water constraints, in many regions, water has become a constraining factor on the energy supply. Power plant operators sometimes don't have access to enough water to build new power generation facilities using conventional designs. And they face environmental constraints on the temperature of cooling water that is discharged into local streams based on limits established to protect fish and ecosystems. Water can also be a constraining factor in extracting energy sources such as in oil and gas production, which can use tremendous amounts of water.

This interdependence between water and energy is called the energy-water nexus. And while the relationship can be mutually constraining, it also presents an opportunity to address both energy and water issues together, because conserving one leads to conservation of the other. Consequently, the way we manage the delicate relationship between the two will have major implications on the future of our energy and water crises.

U.S. Trends

Before the advent of centralized water and mechanical pumping systems, people typically drew water from their local stream, river or well, even in urban areas. The only energy involved was that expended to draw and carry the water. In many rural or poor parts of the world, this is still the case. Breaking from that tradition, the U.S. began developing its centralized drinking water treatment and distribution systems in the early 1900s and now enjoys one of the safest and most abundant public water supplies in the world.

Due to increasing demand, current trends in the United States' water sector suggest that we are moving toward the use of more energy-intensive water: for example, water that has been desalinated in places like Southern California and Florida; or that has been extracted from deep aquifers as in the Colorado Rockies and Great Plains; or that has been transferred between basins, which is done in the Desert Southwest and California.

These trends raise questions about how much energy we use to treat, move and prepare water for end use today, and what lessons learned from the United States' water system can be applied to other regions in the world that are still developing large, centralized water systems and wastewater infrastructure.

Our recent research has examined these questions. To calculate how much energy we use for water on a national scale, we first tallied the energy used in the U.S. to pump water to treatment facilities and treat it. Then we added up water-related home energy use for purposes such as heating water for bathing and cooking. Finally, we included non-household energy used for water in businesses, public facilities, industrial facilities and power plants. We found that annually the U.S. spends 12.3 quadrillion Btu (British thermal units) of energy (about 13 percent of our total annual energy consumption) in one way or another directly on water – much more than we had anticipated.

Energy Used for Water in the U.S.

In the U.S., that 12.3 quadrillion Btu includes uses for everything from extraction through treatment, end uses and disposal. Approximately 30 percent of that energy is consumed exclusively for residential and commercial water heating. Another 30 percent is used to heat and pressurize water for steam injection in industrial processes such as oil refining and chemical manufacturing. Appliances such as dishwashers, washing machines and dryers, which heat water to clean and heat air to dry, represent about 8 percent of water-related energy use.

Although water treatment is envisioned as the poster child of energy spent on water, water treatment uses a relatively small amount of energy compared to that consumed at the point of use. Energy consumed for water treatment by public water utilities represents only 4 percent of the 12.3 quadrillion Btu (equating to 0.5 percent of total 2010 U.S. energy consumption). The remaining 28 percent is spread across multiple uses.

The energy required for water treatment is dictated by the water quality of the source and specific end use. Groundwater typically requires more energy for pumping than surface water, and those energy requirements increase with well depth.

But surface water, which is sometimes degraded from runoff or industrial discharge, often requires more treatment than large groundwater sources, which tend to be cleaner. Once treated, the water is pumped from the treatment facility through the water distribution system to its final end user, where it might be heated, pressurized, pumped or cooled.

Approximately 40 percent of the water that leaves a water treatment plant is returned to the environment through outdoor irrigation or leaks (although this proportion varies a great deal by location and season). Water that is used inside and flushed down the drain (or the toilet) is delivered to a wastewater treatment facility and reconditioned to a cleanliness level that is appropriate for release into the environment. A small fraction of the reconditioned water might be "reclaimed" for nonpotable purposes such as irrigating golf courses or for power plant cooling.

Thermoelectric power producers, agricultural users and many industrial facilities typically extract their own water, as opposed to receiving it from the public water supply. For some activities, raw water quality is sufficient and does not require additional treatment. For other applications, on-site water treatment such as demineralization might be required. Contaminated wastewater from self-supplied water users is still required to be treated to a standard consistent with the EPA's Clean Water Act before being discharged to the environment.

Regional Challenges, Regional Solutions

Despite these broad characterizations and averages regarding national water-related energy use, the United States is a difficult place to generalize. Disparate climates with varying amounts of precipitation and susceptibility to drought affect the availability of surface water and groundwater, which affects pumping depths and distances. Homeowners in Southern California, for example, are likely to receive water that has been pumped hundreds of miles, through two mountain ranges, from the San Joaquin Delta in Northern California. Before the water even reaches its intended customers, it has an energy intensity of about 11 kilowatt-hours (kWh) per 1,000 gallons (though some of that is recovered with in-line turbines as the water flows within pipes back down from the mountain passes). By contrast, customers in Massachusetts, where precipitation and water reservoirs are ample, receive water that has an intensity of about 1.5 kWh per 1,000 gallons, a mere 14 percent of their California counterparts. (However, even in Southern California, end-use activities such as water heating, cooling, pumping and pressurization at the point of use still represent nearly 60 percent of the total energy embedded in water over its entire life cycle.)

The energy consumed by public water and wastewater utilities may currently represent a small slice of water-related energy, but many public water systems around the country are shifting toward more energy-intensive water sources that will likely increase their overall energy use in the future. States like California, Florida and Texas have built desalination facilities that, on average, require about 10 times more energy per unit of water treated than standard surface water treatment operations. Plans for long-distance pipelines to bring water to the drought-stricken West are underway, and although they might temporarily placate water shortages, they have high energetic and financial costs.

In the meantime, droughts in the West, Great Plains and Midwest have increased pressure on over-pumped aquifers, causing water tables to fall and forcing users to draw up from deeper water levels.

Historically, cities and communities grew around available water sources, with the assumption that those sources would last indefinitely. But with new demands on water resources, this planning assumption is challenged. As the water moves to new locations, one question that looms is whether we will move the people to be close to the water or move the water to be close to the people.

There are some bright spots in the United States' water picture. Reclaimed wastewater is being used for purposes such as landscape and golf course irrigation, power plant and industrial cooling, and toilet flushing. It is also being used to replenish aquifers: In Orange County, Calif., falling aquifers have become increasingly susceptible to saltwater intrusion. In response, scientists and engineers designed a system to reinject treated wastewater into the depleted aquifers to create a barrier to keep saltwater from contaminating valuable freshwater resources.

Water conservation has also led to important reductions in demand. Water-efficient cooling technologies for power plants reduce their water demand. Many industries have reduced the water intensity of their supply chains to decrease vulnerability to water shortages. More-efficient irrigation systems, low-flow appliances and xeriscaping – landscaping with native plants that do not require irrigation – also offer water savings for municipalities. Significant savings in irrigation remain for the agricultural sector.

Follow the Leader?

Like the U.S., other countries are moving toward more energy-intensive water via large infrastructure projects intended to deliver clean water from water-rich regions to water-scarce regions. In China, the South-North Water-Transfer Project, slated for completion in 2050, includes plans for three water pipelines of about 300, 700 and 800 miles in length, respectively. Egypt, India, Libya and South Africa, as well as other developing and developed countries, are planning or constructing very large water-supply projects. Similarly, countries such as India, Kuwait, Saudi Arabia, Singapore and the United Arab Emirates have already committed to large desalination facilities to increase potable water supplies. But these projects are very costly and markedly increase the energy consumed to bring water to people around the world.

There is a delicate balance between energy and water resources. Large water infrastructure projects bring water to people who might not otherwise have it, but they also stress the energy infrastructure and impact efforts to move toward alternative sources. However, conservation, reclaimed water projects, and desalination powered with renewable energy could achieve energy, water and climate objectives, simultaneously. For example, Saudi Arabia is building the first commercial-scale solar-powered seawater reverse osmosis desalination plant in the world. Once construction is completed the plant is anticipated to meet the daily water needs of 150,000 people.[5] In the Desert Southwest of the U.S., something similar could be done to turn the vast brackish water resources into freshwater using renewable wind or solar resources.

How the U.S. and other water-constrained countries manage water during the transition from nonrenewable to sustainable sources of energy poses one of the biggest ongoing challenges of the 21st century.

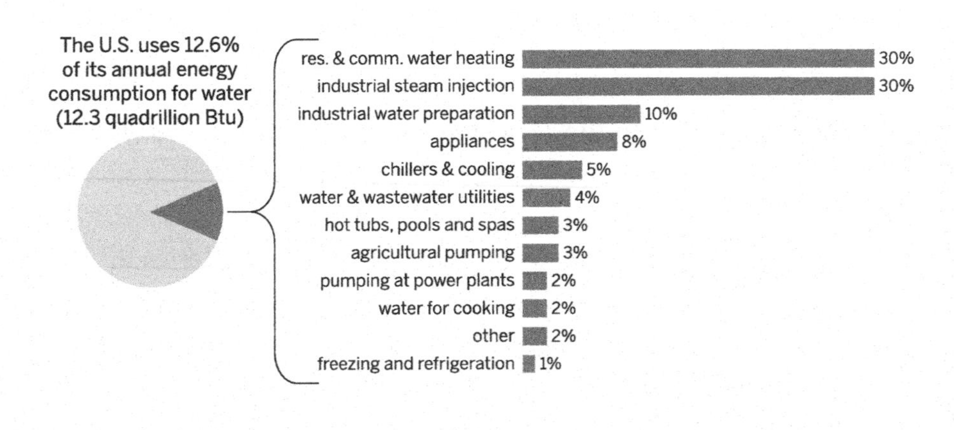

5 According to engineering firm Savener, the facility was commissioned in 2017.

What's the World to Do About Water

POPULAR SCIENCE, JUNE 2014

> *In the end, we can solve the water problem. But we need aggressive conservation that will buy us time while our inventors get to work.*

IN 2007, WHEN MY DAUGHTER WAS 7 YEARS OLD, WE WOULD brush our teeth together every night as part of our daily ritual. To conserve water, we would turn off the faucet after wetting our brushes and turn it back on only to rinse. One night, I didn't turn off the water fast enough to her liking. She turned off the faucet, made an angry face at me, and growled, "Turn off the water, Daddy. The scientists need time."

That statement still resonates with me today, not only because of her precociousness but also because she was exactly right. Kids seem to intuitively get this: We need to conserve our resources to buy ourselves time so scientists can find new solutions to our problems. And this is especially true for water.

Water is on track to be the most important and most contentious resource of the 21st century. It could replace oil as the strategic resource that triggers geopolitical conflict. But with the right solutions, it could also be the one that brings us all together.

In many ways, the 1900s was the century of energy conflict, with great skirmishes triggered and fueled by a quest for petroleum. Now, the dawn of a new energy era is just around the corner – with the prices of solar power plummeting, and distributed generation and energy efficiency on the cusp of taking off – and we can project forward to a time this century when water replaces energy as the next great challenge for humanity. Getting water right could clear the path to a fully liberated, healthy and peaceful civilization.

But water is complicated. First, there's no alternative. There are alternatives to coal and petroleum, but water cannot be replaced. Second, water is intertwined with every other sector of society. Energy production requires water for cooling power plants and fracturing shale. Agriculture needs water for irrigation. Industry and cities use trillions of gallons of water for all sorts of purposes. Third, water demands are growing at the same time supplies are fluctuating.

Climate change is expected to intensify droughts and floods while shifting where water will be and when. The Southwestern U.S. feels that today, with disappearing snowpack and dropping reservoirs.

Thankfully, there are solutions. Some are large, incredibly expensive and energy intensive, such as building cross-continent water-transfer aqueducts, new hydroelectric dams and massive desalination plants. Some are small, such as using micro water harvesters that condense water vapor out of the air. All of them take a while.

In the end, we can solve the water problem. But we need aggressive conservation that will buy us time while our inventors get to work.

Our Water System: What a Waste

THE NEW YORK TIMES, MARCH 2016,
POSTED ON WORLD WATER DAY

> *Putting a sensible price on water to invite investment and encourage conservation, increasing the availability of information and doubling down on innovation can go a long way toward solving it.*

AMERICA HAS A WATER PROBLEM. TO PUT IT SIMPLY, THE NATIONAL network for providing safe, clean water is falling apart.

This state of affairs, which is the focus of a summit meeting on World Water Day at the White House, threatens more than our drinking water supplies. Water is used in every sector of industry, grows our food, affects our health and props up our energy system.

The price of this neglect will be high. In Flint, Mich., the mayor has estimated that it will cost as much as $1.5 billion to fix or replace lead pipes.[6] Overall, repairing our water and wastewater systems could cost $1.3 trillion or more, according to the American Society of Civil Engineers. We need to do this to improve water quality, protect natural ecosystems and ensure a reliable supply for our cities, agriculture and industry.

The problem is a result of many factors, including old, leaky pipes; archaic pricing; and a remarkable lack of data about how much water we use.

In cities across the country, billions of gallons of water disappear every day through leaky pipes. Houston alone lost 22 billion gallons in 2012. As water expert David Sedlak of the University of California, Berkeley, has noted, the water system is facing a double whammy: It has reached the end of its service life just as climate change and population growth have increased its burdens. No wonder the American Society of Civil Engineers gave the nation's drinking water systems a grade of D in 2013.[7]

Wastewater treatment systems are also in serious need of upgrading. Flooding strains treatment plants and sewer systems in many older cities, causing them to discharge untreated sewage whenever rainfall or snowmelt overwhelm them. After Hurricane Sandy, treatment plants in the New York area backed up, with sewage flowing the wrong direction from drainage pipes. The *New York Times* noted that in one neighborhood "a plume of feces and wastewater burst through the street like a geyser."

Droughts also jeopardize water supplies, causing cities in the West to reach farther or dig deeper to get their water. Outside Las Vegas, Lake

---

6 The replacement costs for lead pipes in Flint, Mich., was later determined to be closer to $500 million or more. Replacing lead pipes nationally might require between $45 and 60 billion. Source: https://www.npr.org/2021/05/25/997671845/president-biden-wants-to-replace-all-lead-pipes-flint-has-lessons-to-share [Accessed November 12, 2022]

7 For the American Society of Civil Engineers 2021 infrastructure report card, the rating was slightly improved, earning a C- instead of a D.

Mead, fed by the Colorado River, was recently measured at 39 percent of capacity.[8]

These problems are compounded by an antiquated system of regulations, dysfunctional water markets, policies that encourage over-pumping, and contracts that discourage conservation by requiring customers to pay for water they don't use. These approaches depress investment and inhibit innovation.

To fix our water systems, we need prices that lead to more rational water use and invite needed investment, data to track water resources and usage, and much more research and development.

Take prices, for example. Water prices should rise or fall according to supply and demand. The idea that the price should be the same in the dry season (when supplies are low and demand for irrigation is high) as the wet season (when supplies are high and demand is low) is nonsense.

Water utilities should take a page out of the energy sector's playbook. Electric utilities had been plagued for decades by many of the same difficulties. But now they are moving toward time-of-use pricing, with prices rising when demand is up, and inverted block pricing, where prices increase with consumption. Allowing these price shifts would change user behavior. Higher prices would encourage conservation and new technologies.

Regulations can ensure that the first few gallons per person per day are cheap or free, with escalating costs beyond that. Water for necessities such as drinking, cooking and hygiene should be affordable. Beyond that, water for lawns, filling swimming pools, washing cars and other uses should be more expensive.

We also have to fix our data gaps. We are operating blind. Compared to sectors like energy, where robust statistics on prices,

---

8 According to NASA's Earth Observatory, by July 2022, the water in Lake Mead had dropped to 27% of capacity.

production and consumption are generated weekly, key information on water use and supply is missing or published only every few years.

We should increase the federal budgets for water monitoring. Establishing a Water Information Administration, just as the Department of Energy has an Energy Information Administration, to collect, curate and maintain up-to-date, publicly available water data would inform policymakers and the markets.

Congress should also significantly increase support for water research and development, making sure to include the private sector as a partner.

We need breakthroughs in water treatment technology that would enable larger-scale recycling and reuse of treated water, desalination, and aquifer storage and recovery. These improvements range from the mundane – better pumps and home appliances – to advanced nanomaterials for energy-efficient water treatment.

The water sector's risk-averse culture has resisted innovation. Higher prices and government-backed research and development could help prompt a wave of innovation and investment. This is what happened with hydraulic fracturing and horizontal drilling, two technologies advanced through government research that kicked off the shale boom.

The water problem is daunting. But putting a sensible price on water to invite investment and encourage conservation, increasing the availability of information and doubling down on innovation can go a long way toward solving it.

The Water Trade

MECHANICAL ENGINEERING, NOVEMBER 2016

> *In a world of declining freshwater availability, we are exchanging energy for clean water to meet the needs of thirsty billions.*

THE WORD "TANTALIZE" HAS ITS ROOTS IN A WATER-BASED LEGEND. THE GREEK GODS PUNISHED TANTALUS, A SON OF ZEUS, BY giving him a great thirst and forcing him to stand in a pool of water that always recedes as he leans down to take a drink.

Such a myth feels like a fitting parable for humankind's relationship with abundant water resources that seem to be forever beyond our reach. In fact, it is this inconvenience that drives much of our energy investments for water: we spend significant sums of energy moving, treating, or storing water so that it is available in the form, location, and time we want it. While those energy investments overcome the limits of water's tantalizingly distant location, billions of people still remain without clean, accessible water.

What's more, demand for energy and water has been growing faster than population, driven by economic growth on top of the population growth. Affluent people eat more meat and consume more electricity – two activities that use water. With many water withdrawals coming from nonrenewable resources, the trends for greater consumption will trigger water shortages unless something changes.

By 2005, at least half of Saudi Arabia's fossil (nonrenewable) water reserves had been consumed in the previous two decades. Significant declines have also been observed in the Ogallala Aquifer under the Great Plains of the United States, spanning eight states from South Dakota to Texas. Water tables in Texas have lowered by as much as 234 feet, while the average drop across the entire aquifer was 14 feet. Storage of water fell from 3.2 billion to 2.9 billion acre-feet.

Overall, water availability is declining globally. Available water dropped from 17,000 cubic meters per person in 1950 to 7,000 cubic meters per person in 2000. Water stress occurs between 1,000 and 1,700 cubic meters, and a water crisis occurs at less than 1,000 cubic meters. Notably, counties such as Qatar, Libya, and Israel are well below 400 cubic meters per person, and even the "green and pleasant" United Kingdom only has 1,222 cubic meters.

All these datasets point toward a conclusion that water stress is increasing. High-profile research published in *Nature* has concluded that nearly 80 percent of the global population endures high levels of threat to water security. To compensate for the decline in water availability, we are moving toward more energy-intensive water. This relationship is just one aspect of the energy-water nexus: the dependency of water availability on energy inputs, and the dependency of power generation on available water. The increased energy intensity of water has several different components, including stricter water/wastewater treatment standards, deeper aquifer production, long-haul pipelines, and desalination. Each of those elements is more energy intensive than conventional piped water today, and seems to be a more common option moving forward.

Cleaner

As societies become wealthier, their concerns shift from focusing on economic growth to protecting the environment. Protecting drinking water quality and preserving the ecosystem from the discharge of water treatment plants are two important pieces of that trend.

But water and wastewater treatment require nontrivial amounts of energy, and advanced treatment methods to meet stricter standards are more energy intensive than treatment for lower standards. For example, advanced treatment systems for wastewater with nitrification require about twice as much energy as trickling filter systems. As we tighten the standards for water and wastewater treatment, we are essentially edging toward increases in energy consumption. While new treatment technologies and methods become more efficient over time after their initial implementation, the standards tighten in parallel. How this balances out is unclear.

At the same time, the water coming into water and wastewater treatment plants is getting more polluted with time. As population grows, there are more discharges into the waterways. Those discharges contain constituents that weren't always there in such high concentrations. For example, there have been growing concerns about pharmaceuticals (including birth-control pills and pain pills) in sewage streams, which are difficult for wastewater treatment plants to remove. Doing so requires new equipment and ongoing investments of energy.

In an ironic example of the energy-water nexus, some of our energy choices create water-quality impacts that require additional energy to treat. For example, increased biofuels production from Midwest corn is expected to cause additional runoff of nitrogen-based fertilizers and other pollution that will require more energy to remove.

Also, the wastewater streams from hydraulic fracturing of shales to produce oil and gas contain much higher levels of total dissolved solids than most wastewater treatment plants can handle. That means more energy must be spent in one of several ways: on trucking that wastewater to disposal sites or specialized industrial wastewater treatment facilities that might be far away (something that happens rarely); for subsurface sequestration; for on-site treatment to recycle and reuse the water in subsequent wells; or on new equipment at the wastewater treatment plant to treat those streams. Even that new equipment is sure to require energy.

Farther

We are also contemplating moving water farther from its source to the end user. Long-haul pipelines and interbasin transfer, which is moving water from one river basin to another, are common proposals to solve the crisis of declining local water supplies. While the idea of aqueducts has been around for thousands of years, the scale, length and volumes of water that are moved are growing.

Some of the classic water-transfer systems include the State Water Project in California, which is the state's largest electricity user because it must pump the water over mountains. (It also captures a lot of energy when the water flows back downhill through in-line hydroelectric turbines coupled with chutes.) Maui is another example. The island has an incredible series of hand-cut water channels that circle its two volcanoes, moving water miles from the wet portion of the island – one of the wettest places in the United States – to the dry inland plains where farming occurs. This system operates by gravity and generates electricity along the way.

Moving forward, as water tables fall and surface sources dry up, municipalities are more likely to consider the cost of expensive and far-flung water gathering systems that pull water to a city from deeper in the ground or farther away. These long-haul systems will generally not be gravity-fed and will require a lot of energy. Plus, they will impact the ecosystem as water from one basin is moved to another, both in terms of loss of water in one watershed and the potential for invasive species in the other.

Perhaps the most ambitious water project in the world is the South-North Transfer Project in China (also known as the South-North Water Diversion Project). The scale, scope and ambition of the project is reminiscent of U.S. water planners who have dreamed for decades of diverting the Yukon River in Alaska or the Missouri River in the Great Plains to the American Southwest, so that the deserts would bloom with flowers and fruit trees. This project essentially aims to move major southern rivers – the Yangtze and Han – across the country to the Yellow and Hai rivers. The industrialized north is relatively water poor, whereas the southern part of China is relatively water rich. The total estimated flow for the Chinese endeavor is projected to divert 44.8 billion cubic meters per year from the south more than a thousand miles to the north, at a total cost estimated to be $62 billion.

Not to be left out, India is also building its own long-haul water pipeline and other countries are doing the same.

Fresher

Another of the key trends to watch is how many municipalities are turning to desalination as a solution for water-supply issues. In 2013, more than 17,000 desalination plants were already installed worldwide, providing approximately 21 billion gallons per day of freshwater. With a blistering pace of growth, that capacity is projected to keep expanding quickly. More than three-fourths of new capacity will be for desalinating seawater, with the rest from brackish groundwater or salty rivers.

While thermal desalination (using heat) represents about 25 percent of the installed capacity by 2010, it represents a shrinking share of new installations as builders seek the less energy-intensive reverse osmosis membrane-based system. Even with the lower-energy approach, desalination is still an order of magnitude more energy intensive than traditional freshwater treatment and distribution. Desalination is capital intensive too: The annual global desalination market exceeds $10 billion.

Growth in desalination is particularly rapid in energy-rich, water-poor parts of the world, such as the Middle East, northern Africa and Australia. After a severe drought that lasted several years, water-strapped Israel famously turned to the sea for its water, rapidly building a handful of desalination plants to produce about 200 billion gallons of freshwater annually by desalting water from the Mediterranean.

Rapid desalination growth is also occurring in China, where booming industrial activity is straining water supplies that serve the world's largest population. It is also popping up in locations such as London, where a new desalination plant was very controversial and became a big part of several mayoral campaigns.

Despite its relative water wealth, the United States is the world's second-largest market for desalination, trailing only Saudi Arabia. This phenomenon is partly the result of the unequal distribution of water resources across the United States. And, as a wealthy country, the water consumption per capita is quite high and the money to finance large-scale infrastructure projects is available. Projects are under consideration for seawater reverse osmosis in coastal states such as California, Texas and Florida. And projects are under development to serve inland communities that sit atop large brackish aquifers, as in Texas, Arizona and New Mexico.

The two most energy-intensive options – desalination and long-haul transfer – can also be combined to create an even larger energy requirement for water. Natural water flows occur by gravity, but for seawater desalination, the opposite is true. By definition, coastal waters are at sea level, so moving the water inland requires pumping water uphill. One such desalination project under development in the United States is a coastal facility along the Gulf of Mexico that is designed to provide freshwater for San Antonio, Texas. That means the water would be moved nearly 150 miles inland, increasing in elevation nearly 775 feet.

While trading energy for water makes a lot of sense in places like the Middle East or Libya, where there is an abundance of energy and a scarcity of freshwater, that trade is not obviously a good value in places like the United Kingdom or the United States, where other cost-effective options such as water conservation, graywater capture and water reuse might be available.

In the end, the most important innovation we need is a new way of thinking about energy and water so that we make better decisions about those precious resources: holistic thinking that recognizes these resources as interconnected, and a systems-level approach that acknowledges how one change in one state to a water system can impact an energy system five states away.

Most important, we need long-range thinking because our energy and water decisions last decades to centuries, so it's imperative that we get them right.

Time for Another Giant Leap for Mankind

ISSUES IN SCIENCE AND TECHNOLOGY, SPRING 2012

By Sheril R. Kirshenbaum and Michael E. Webber

> *Just imagine what we would have accomplished by now had we devoted the same attention to looking for water on Earth as looking for water on the moon.*

IN MAY 1961, PRESIDENT JOHN F. KENNEDY ANNOUNCED A BOLD priority for the United States. He memorably urged the nation to send a man to the moon by 1970: "No single space project in this period will be more impressive to humankind, or more important for the long-range exploration of space; and none will be so difficult or expensive to accomplish."

Kennedy and his administration recognized that the United States risked losing the space race to the Soviets during the Cold War. With tremendous federal support from Congress, it took just eight years before Neil Armstrong left dusty footprints behind on the moon.

There were other significant challenges he addressed as well that we don't hear as much about. Just one month earlier, the president pressed the country to overcome another serious task:

"If we could ever competitively – at a cheap rate – get freshwater from saltwater, that would be in the long-range interest of humanity and would really dwarf any other scientific accomplishment."

Kennedy understood that improving desalination technologies would raise men and women from lives of poverty and improve human health globally. Unfortunately, that quote failed to make many history books.

Why should water be a great national priority? Even though the planet is covered in water, only 2.5 percent of it is fresh and two-thirds of that is frozen. This doesn't leave much for the estimated 10.1 billion people on Earth by 2100. Water consumption continues to increase at a faster rate than population growth. Today more than a billion people do not have access to clean drinking water, and the United Nations estimates that unsanitary water leads to more than 2 million deaths every year. Waterborne illnesses are associated with 80 percent of disease and mortality in developing nations; sadly, the majority of the victims are children.

Since Kennedy's 1961 call to arms, neither desalination nor the larger issue of the nation's water infrastructure has received much public attention or regular directed federal support. Water R&D has not been a consistent priority, and investment has endured erratic boom and bust cycles.

During the 1960s and 1970s, the U.S. government cumulatively spent more than $1 billion (in nominal dollars) on desalination R&D alone. The Water Resources Research Act of 1964 led to the creation of the Office of Water Research and Technology in the Department of the Interior in 1974 to promote water resources management. It also helped to establish water research institutes at universities and colleges. Three years later, the Water Research and Conservation Act authorized $40 million for demonstration-scale desalination plants. The following year, the Water Research and Development Act extended funding through 1980. But the Office of Water Research and Technology did not last. Just eight years after it opened, the Reagan administration abolished it, distributing authority over water programs among various agencies.

Congressional appropriations continue to be provided annually to fund water infrastructure, but it's been 16 years since authorizing legislation was enacted to set drinking water policy. For wastewater, it has been a quarter century. Complicating matters further, because water programs are spread across a host of agencies and departments, it's extremely difficult to track government R&D spending on water.

However, with the help of the National Academy of Sciences and the American Association for the Advancement of Science, we have been able to assemble a first-cut estimate for water R&D over time [in fiscal year (FY) 2011 dollars]. Although the details might be a little fuzzy, the overall picture is clear: at a level of about $1 billion per year, Water R&D has been woefully neglected.

How can the nation expect to meet the looming water challenges when spending has not even been reliably tracked for the past 50 years?

If it's true that we cannot improve what we do not measure, then the fact that water R&D hasn't been carefully tracked is a sign that we're not taking it seriously. It's no wonder that water treatment technologies have evolved so slowly, that water infrastructure leaks so abundantly, and that water quality is at risk from a variety of societal activities and policy actions. Despite decades of building the greatest innovation and R&D system the world has ever seen, progress in water innovations seems halting and stunted, especially when compared with the advances that occurred in parallel for information technology, energy, health care or just about any other sector critical to society.

Just imagine what we would have accomplished by now had we devoted the same attention to looking for water on Earth as looking for water on the moon.

Today, even the youngest Americans can quote Neil Armstrong. We celebrate the space program as one giant leap for humankind. Now it's time to take a second great leap by doing something even greater for humanity: investing in water research.

Energy and Thanksgiving

RADIO STATION KUT, NOVEMBER 13, 2013

> *We spend ten times as much energy on our food as the food itself contains.*

THANKSGIVING IS HERE, AND THAT MEANS TWO THINGS: A LOT OF travel and a lot of food. Both of those require a lot of energy. The travel itself consumes a lot of energy.

Throughout the year, transportation is responsible for 28 percent of our energy consumption. And there is a nontrivial bump right around Thanksgiving time. According to *USA Today*, more than 25 million people in the United States are expected to fly for the

Thanksgiving holiday. That's more than 2 million people per day for the 12-day holiday period.[9]

Those airplanes consume a lot of jet fuel. The bad news for travelers is that those planes will be full. The good news from an energy perspective is that full planes get better mileage than half-full planes. In fact, full planes get something like 80 to 100 miles per gallon, which is pretty good, except that we tend to fly when we're traveling long distances, so the total energy consumption adds up quickly. In addition to all those flyers, tens of millions of other Americans will hit the roads.

Another part of Thanksgiving's energy picture is how much energy is used for food. Across the nation, about 10 percent of our annual energy consumption is embedded in our food system. We consume about 100 quads (quadrillion Btu) of energy each year; that means we spend about 10 quads of energy on our food system. But that food only contains about 1 quad of nutritional energy, which our bodies use for our various muscles and cellular functions, for example to stay alive and for exercise.

That means we spend ten times as much energy on our food as the food itself contains.

Those energy investments are for things like growing, harvesting, processing, transporting, packaging, preparing, cooking and disposing of food. It includes the energy in the fertilizers (which we make from natural gas), the pesticides (which we make from petroleum), the diesel for the farm equipment and trucks that collect and move the food, the natural gas for cooking the food, the electricity to heat the water we use to wash our dishes, and so on.

Historically, the amount of energy we spend on food has been a good news story: It means that compared with most countries, we have plenty of food and it's mostly affordable. And it means we have interesting outcomes, such as dishes that have traveled the world before arriving

9 Thanksgiving 2022 numbers were similar to 2013 after rising to a record high in 2019 and then dropping in 2020 and 2021 due to the COVID-19 pandemic.

at our plate or fruits that are never out of season because we import them from different hemispheres throughout the year.

In an era of increasing concerns about food security and environmental impacts of energy and agriculture, those energy investments and global food systems strike many people as a luxurious indulgence we can no longer afford.

As we grapple with this conundrum, there are two things to keep in mind: Meats and dairy are more energy intensive than plant-based foods, and we throw away a lot of food.

One consequence of living in a rich country with an abundance of food is that we waste a lot of food. Estimates vary, but anywhere from 25 to 50 percent of our edible food is thrown away. That by itself is tragic enough, especially considering how many people in the world are hungry. But if we remember how much energy is embedded in that food, the story becomes worse still. At least 2 percent of our national energy consumption is embedded in the perfectly good food we throw away. To put it in context: We throw away more energy in the food we discard than Switzerland uses in an entire year for all purposes.

In an era where we care about the energy requirements and environmental impacts of food, reducing our meat intake and our food waste is a pretty straightforward way to go.

Bringing all of these together: If you want to have an energy-efficient Thanksgiving and you can't reduce the number of miles you travel and you don't want to reduce the amount of meat you eat, then reducing the amount of food you throw away might be a good step in the right direction.

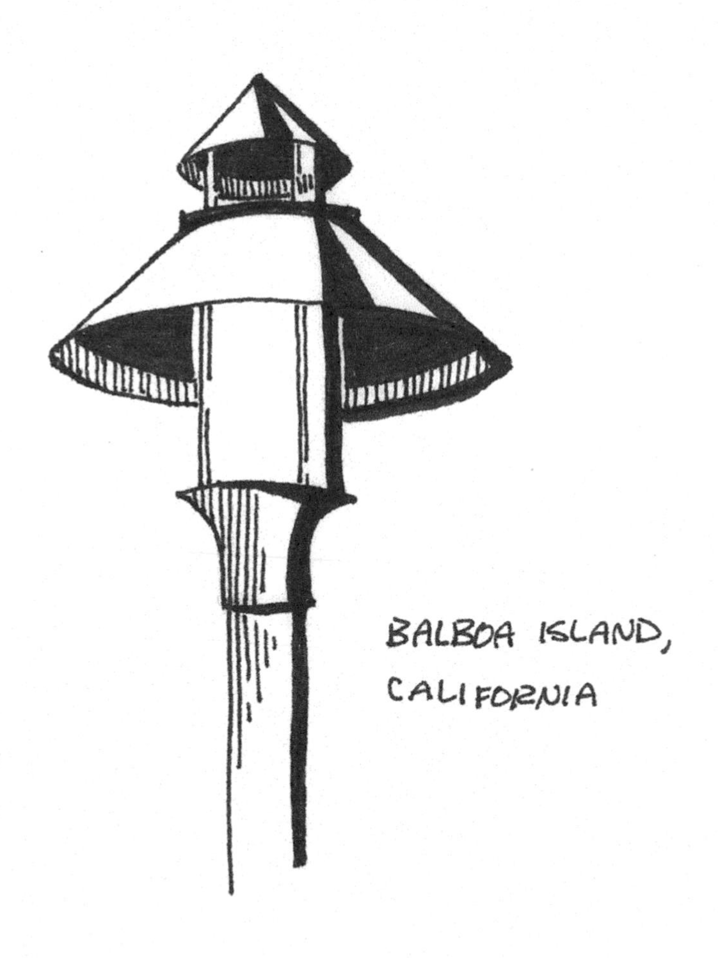
BALBOA ISLAND,
CALIFORNIA

Chapter 2

Physiological Needs II: Garbage In, Garbage Out

Trash-To-Treasure: Turning Nonrecycled Waste Into Low-Carbon Fuel

EARTH MAGAZINE, AUGUST 2012

By Alex C. Breckel, John R. Fyffe and Michael E. Webber

Could garbage one day replace coal at power plants?

AMERICANS PRODUCE MORE THAN FOUR POUNDS OF TRASH PER person per day, amounting to 20 percent of the world's waste. Although recycling rates have increased over the past few decades – out of the 4.9 pounds of trash (per capita) that we produce in the U.S. each day, we compost or recycle about 1.5 pounds and incinerate a little over 0.5 pounds – more than 50 percent of our waste still ends up buried in landfills. Now, through the use of a novel energy recovery technique, we could reduce the amount that is sent to landfills by about 10 percent and produce a fuel that is relatively clean and more energy dense than coal.

Materials that were considered garbage for generations are now being recognized for what they still offer after their useful life – valuable energy resources capable of solving multiple problems at once. Trash can become treasure.

Producing energy from trash is known as a "waste-to-energy" option. Several such options have existed for many years and are in extensive use throughout Europe and limited use in the United States. One of the more exciting options that has been proposed within the last decade is to convert waste into solid recovered fuels, or SRFs. SRFs are engineered blends of nonrecycled waste condensed into fuel pellets or cubes.

This opportunity is particularly appealing for plastics that are hard to recycle, decompose slowly in landfills, and have higher energy density than coal – baby diapers, for example. Although diapers serve an important purpose during the normal product cycle, once they have been used, they are too complex to economically recycle, and as a consequence they are typically discarded, where they remain for what we expect will be thousands of years. Although they are almost certain to be an interesting archaeological find many centuries from now, today they could make a great fuel. That invites the broader question of how many other nonrecycled plastics could be turned into fuels instead of wasted in landfills.

It's a question our team at the University of Texas at Austin decided to examine. Wanting to see the real-world capabilities of SRFs, we worked with partners in the plastics, recycling, pelletizing and cement industries to conduct an experiment in which nonrecycled waste was processed into SRF pellets. This SRF was then burned in a cement kiln, a process that normally uses a lot of coal and other fossil fuels; the SRFs replaced a fraction of these fuels. The experiment's results demonstrate the potential for SRFs to displace fossil fuels for energy, but they also reveal some initial hurdles that must be overcome before SRFs can become a large-scale reality.

The Hierarchy of Trash

The U.S. Environmental Protection Agency has created a waste management hierarchy to highlight the ideal method of dealing with solid waste. The old adage of "reduce, reuse, recycle" still holds true as the most preferred path to manage our waste, but not all recyclables are clean enough or well-enough separated to be economically recovered.

For example, many regions in the U.S. have started switching to single-stream recycling where all recyclable materials are mixed in a single bin at home. This trend made it easier for homeowners and businesses to recycle their waste. The commingled recyclables are taken to a materials recovery facility where the commodities are separated and sold to recycling facilities or sent to landfills. Although single-stream recycling has contributed to the increase in overall recycling rates in the U.S., it unfortunately makes it easier for paper or plastics to become contaminated by other materials like food and liquids in the recycling stream, making them unfit for sale to recyclers.

Because of contamination and imperfect sorting, between 5 and 25 percent of a materials recovery facility's incoming recyclables are discarded and sent to landfills. The waste, called "residue" in the waste management business, is a valuable mixture of paper and plastics that is currently lost to the ground. And this is where SRFs come in, taking advantage of an energy-dense waste stream that can be recovered to provide economic, environmental and resource conservation benefits.

Our Waste-to-Energy Options

Currently, there are several viable waste-to-energy options, ranging from low-tech to modern.

Mass incineration – the simple process of burning trash to make heat or electricity – has been used for decades. Furthermore, this technology has made progress recently in alleviating many of the environmental concerns over the emissions from burning trash. Unfortunately, incineration is not always complementary to recycling because they sometimes draw on some of the same materials. It is also very expensive to build new highly specialized incinerators with the necessary advanced emissions control systems required by environmental regulations. Therefore, its prospects for expanded operation are hard to predict.

Another potential solution for recovering energy from waste is pyrolysis, in which plastics are thermally processed to molecularly break them down into the building blocks of fuels that can then be processed into gases, oils or even high-quality liquid fuels that could be used in place of gasoline. There is a strong desire for alternative energy sources to displace petroleum in liquid fuels markets, making pyrolysis an appealing option. Several companies in the U.S. and Europe have recently commercialized pyrolysis techniques, and many more are in the initial testing phases. However, the incoming material typically must be high-quality homogeneous streams of plastics, making it an unlikely solution (with today's technology) for the mixed residue stream coming from materials recovery and recycling facilities.

Because of these challenges, SRFs – which integrate the energy recovery piece of the puzzle with the reuse and recycle pieces and enable us to alleviate environmental problems while still recovering the recyclable materials – look to be like a pretty good bet. Producing SRFs from waste is not entirely new: Several American and European companies are already producing quality pelletized fuel from trash. The novel idea that we pursued was to use the materials from materials recovery facilities that cannot be economically recycled as a feedstock to produce SRFs.

The Fuel

SRFs can be created by selectively mixing and shredding a blend of plastic and paper materials and then densifying that blend into a solid pellet form. What makes residue-derived SRFs such adaptable and promising fuels is largely that the paper and plastic content have relatively high energy densities and are abundant in the waste stream. Depending on the raw materials used, the energy content of SRFs can be significantly higher than most types of coal. Consequently, SRFs can potentially be integrated into processes that consume large amounts of coal, such as cement manufacturing, or can be co-fired at current coal-fired power plants without distorting the overall energy balances.

And for the most part, combustion of plastics is cleaner than many people think. Most plastics produced in the U.S. are created by tying together building blocks of hydrocarbon polymers, which are composed of hydrogen and carbon atoms, much like the fossil fuels from which they are derived. Most plastics found in consumer products, ranging from deli wrappers to diapers, can burn as clean as, and sometimes cleaner than, natural gas.

Of course, not all plastics are well suited to combustion. Some additives used in the production of plastic or mixed in the final product (such as chlorine) can be harmful to the environment if improperly combusted or emitted without scrubbing. Thus, attention must be paid to ensure that the incoming materials used to produce SRFs are of suitable quality, which in most cases means nothing more than sorting out problematic materials before the production process begins.

Could It Replace Coal?

Coal is SRFs' closest fossil fuel analog in terms of fuel content and handling characteristics. U.S. consumers use nearly a billion tons of this nonrenewable fossil fuel every year for power production, resulting in 2-plus billion megawatt-hours of power—and 2-plus billion metric tons of carbon dioxide emissions per year.[10] Another large end user is the cement industry, which consumes 10 million tons of coal annually and creates more than 80 million tons of carbon dioxide from the combustion of fossil fuels and the chemical reactions used to make cement. At the same time, the U.S. generates more than 290 million tons of municipal solid waste annually,

10 By 2022, these numbers had dropped approximately in half, with annual coal consumption at a little over 500 million tons to generate about 1 billion megawatt-hours of electricity and releasing about 1 billion metric tons of carbon dioxide.

44 million tons of which are nonrecycled plastics and paper products – sufficient to cover the needs of the cement industry.

America's appetite for energy and SRFs' ability to displace coal create a synergistic solution. SRFs derived from the residue of recycling facilities can conserve finite fossil fuel resources by creating a domestic alternative to coal, all the while diverting more waste from landfills and reducing greenhouse gas production. Meanwhile, using SRFs instead of coal would also eliminate some of the other undesirable byproducts of coal production and consumption, including indirect emissions from transporting the fuel across the country (waste streams are usually transported much shorter distances), water risks caused by runoff from mines, and land disturbances caused by surface mining.

SRFs in the Real World

To test the real-world capabilities of residue-derived SRFs, our research team at UT-Austin (together with partners in the waste and cement sectors) conducted a large-scale test burn and analysis of SRFs co-fired with fossil fuels in a cement kiln.

Seventy-five tons of residue from a materials recovery facility in Virginia were gathered and combined with post-industrial waste products such as scrap plastics from a manufacturing plant. The final product, a blend of residue and post-industrial waste in a 60:40 ratio, contained mostly plastics and paper products. A fuel processing facility in Arkansas used this waste to create 130 tons of SRFs in the form of pellets, which we then burned in the precalciner portion of a cement kiln in Texas. The precalciner is a special combustion chamber of a cement kiln that serves to preheat and decarbonate raw materials before entering the main kiln.

Our experimental results showed that the SRFs had a predictable energy content of about 12,500 British thermal units per pound (25 million Btu per ton). Bituminous coal, the type normally used at this particular cement kiln, has almost exactly the same energy density, leading to a nearly one-to-one displacement opportunity. The SRFs produced for our experiment were also 40 percent more energy dense than sub-bituminous coals and 80 percent more so than lignite.

When the whole production, transportation and combustion life cycle of the SRFs is considered, large fossil fuel energy savings can be realized. Our experiment lasted a few days. Extrapolating the fossil fuel displacement rate of one ton per hour that we used in our experimental demonstration over an entire year, SRFs would reduce total fossil fuel energy use by 6 percent annually in the cement kiln. This reduction equates to about 9,000 tons of coal, enough to provide electricity to 1,500 average U.S. homes for a year. Likewise, under this scenario, carbon dioxide reductions of 14,000 tons per year are possible, largely from reduced landfill gas production. This decrease is comparable to removing 2,800 cars from the road. And that's just one cement kiln.

The EPA estimates that 94 million tons of waste flows through U.S. materials recovery facilities each day. When the magnitude of the resultant residue stream is considered, the potential for energy savings and greenhouse gas reductions is immense. The amount of SRF production that could be realized is enough to power nearly a million homes and reduce carbon dioxide emissions by 5 million tons, equivalent to removing 1.3 million cars from the road.

The Long Road Ahead

As successful as our experiment was at demonstrating the benefit of SRFs, it also revealed how much remains to be done before the full potential of residue-derived SRFs can be achieved.

First, new procedures must be established to thoroughly examine residue streams prior to use. Residue from materials recovery facilities is a heterogeneous and relatively unpredictable mixture of waste. Although most materials that end up in the residue stream are suitable for combustion, others can compromise the quality of the SRFs by reducing the handling characteristics, decreasing energy density, or producing undesired emissions when combusted.

Second, more large-scale and long-term tests need to be conducted to develop a full understanding of the challenges and cost-benefit comparison of producing SRFs from materials recovery facility residue. For example, regional and seasonal variation in materials recovery facility residue composition and availability can impact the economics and technology used to produce SRFs, so they need to be well understood before bringing this technology to market. In addition, higher feed rates should be tested to determine the technical limits and further assess the potential for fossil fuel displacement.

Third, some regulatory policies should be reconsidered before SRFs become big business. Currently there is no consensus on how to handle SRFs in the regulatory realm. In a policy context, the term "waste-to-energy" is used as a catchall for many such technologies, including SRFs. And across the U.S., waste-to-energy techniques have mixed support among states. Some states support the technologies as a means of renewable energy generation, whereas others reject them, and most do not address them at all. Wisconsin currently addresses densified fuel pellets – a term that encompasses SRFs and other less refined waste-derived fuel pellets – directly in the state's renewable portfolio standard, which is a positive sign.

This heterogeneous policy landscape hampers interstate business and can be a roadblock to investors trying to seize on new waste-to-energy opportunities. The lack of policies specifically addressing emerging technologies such as SRFs and plastics-to-fuels suggests an information gap between technology developers and policymakers. Fortunately, trends in updating renewable portfolio standards to include waste-to-energy facilities and alternative conversion processes as renewable technologies will lead to a better business environment for companies pursuing energy recovery from solid waste.

Despite technical, social, political and economic hurdles, harnessing the energy content of nonrecycled plastics and papers derived from materials recovery facility residue provides many benefits while complementing regional recycling efforts. Displaced fossil fuels, landfill avoidance and reduced greenhouse gas emissions are just some of the advantages offered by SRF production. As recycling rates continue to increase and SRF production techniques are further refined, residue-derived SRFs will be an important resource to consider as one solution to concerns about America's long-term energy usage, resource conservation and waste management. The materials that even the recycling industry once considered trash could become a national treasure.

Tapping the Trash: Transforming Costly Wastes Into Valuable Resources Can Make Cities Highly Efficient

SCIENTIFIC AMERICAN, JULY 2017

> *"Waste is what's left when you run out of imagination." –David Scott*

ON DEC. 20, 2015, A MOUNTAIN OF URBAN REFUSE COLLAPSED IN Shenzhen, China, killing at least 69 people and destroying dozens of buildings. The disaster brought to life the towers of waste depicted in the 2008 dystopian children's movie WALL-E, which portrayed the horrible yet real idea that our trash could pile up uncontrollably, squeezing us out of our habitat. A powerful way to transform an existing city into a sustainable one – a city that preserves the earth rather than ruining it – is to reduce all the waste streams and then use what remains as a resource. Waste from one process becomes raw material for another.

Many people continue to migrate to urban centers worldwide, which puts cities in a prime position to solve global resource problems. Mayors are taking more responsibility for designing solutions simply because they have to, especially in countries where national enthusiasm for tackling environmental issues has cooled off. International climate agreements forged in Paris in December 2015 also acknowledged a central role for cities. More than 1,000 mayors flocked to the French capital during the talks to share their pledges to reduce emissions. Changing building codes and investing in energy efficiency are just two starting points that many city leaders said they could initiate much more quickly than national governments.

It makes sense for cities to step up. Some of them – New York City, Mexico City, Beijing – house more people than entire countries do. And urban landscapes are where the challenges of managing our lives come crashing together in concentrated form. Cities can lead because they can quickly scale up solutions and because they are living laboratories for improving quality of life without using up Earth's resources, polluting its air and water, and harming human health in the process. Cities are rife with wasted energy, wasted carbon dioxide, wasted food, wasted water, wasted space and wasted time. Reducing each waste stream and managing it as a resource – rather than a cost – can solve multiple problems simultaneously, creating a more sustainable future for billions of people.

Pollution as a Solution

Lessons about waste abound in history. In 1855, John Snow, a London doctor, deduced that terrible cholera outbreaks struck London in 1848 and 1854 because public water wells were contaminated by sewage. Building sewers was an obvious solution, but political leaders rejected Snow's findings because his ideas did not fit prevailing ideologies and because the actions were deemed too expensive. Similar rejection is offered for today's climate scientists, who tell us that our waste is killing us, though in a much slower and less direct fashion, and that fixing the problem will require significant investments in new infrastructure. Snow was later vindicated as a hero (perhaps the same fate awaits our present-day climate scientists) after new leaders created ambitious public works projects to cram 1,200 miles of sewers into a crowded city of 3 million people, ending the cholera problem. The work also created the lovely river embankments that still stand as a key piece of London's urban environs and along which many people stroll.

Today just flushing the waste away is not enough, however. After we reduce it, we should close the loop and use the remainder again. First, limit waste, then put it to work.

This new thinking begins by redefining our concept of pollution. Raj Bhattarai, a well-known engineer at the municipal water utility in

Austin, Texas, taught me a new definition for pollution: resources out of place. Substances are harmful if they are in the wrong place: our bodies, the air, the water. But in the right place, they are useful. For example, instead of our sending solid waste to a landfill and paying the bill, it can be incinerated to generate electricity. And the sewage for a million-person community can be mined for millions of dollars of gold and other precious metals annually for use in local manufacturing.

This idea fits with the larger concept of the so-called circular economy – where society's different actions and processes feed into one another beneficially. As David Scott, a forward-thinking energy expert, taught me: "Waste is what's left when you run out of imagination."

Less Is More

One obvious place to start reducing waste is leaky water pipes. A staggering 10 to 40 percent of a city's water is typically lost in pipes. And because the municipality has cleaned that water and powered pumps to move it, the leaks throw away energy too.

Energy consumption itself is incredibly wasteful. More than half the energy a city consumes is released as waste heat from smokestacks, tailpipes, and the backs of heaters, air conditioners and appliances. Making all that equipment more efficient reduces how much energy we need to produce, distribute and clean up.

Refuse is another waste stream to consolidate. The U.S. generates nearly five pounds of solid waste per person every day. Despite efforts to compost, recycle or incinerate some of it, about half is still dumped in landfills. Reducing packaging is one way to lessen this volume while also generating other benefits. Big retailers such as Walmart, for example, have found that reducing packaging results in fewer trucks needed for shipping and more shelf space to display goods.

Wasted food is its own heart-wrenching issue. Despite famine and food scarcity in many places globally, Americans throw away 25 to 50 percent of their edible food. Food requires vast amounts of energy, land and water to grow, produce, store, prepare, cook and dispose – so wasted food leaves a significant imprint. Initiatives that have popped up in the

U.K., and the U.S., such as the *I Value Food* campaign, are a start toward solving this vital issue.

Putting Waste to Work

Once cities reduce waste streams, they should use waste from one urban process as a resource for another. This arrangement is rare, but compelling projects are rising. Modern waste-to-energy systems, such as one in Zurich, burn trash cleanly, and some, including one in Palm Beach, Fla., recover more than 95 percent of the metals in the gritty ash that is left by the combustion. Rural villages, such as Jühnde in Germany, create enough biogas from cattle and pig manure to heat or power a large portion of their homes. My research group at UT-Austin has demonstrated that a cement plant in New Braunfels, Texas, can burn fuel pellets made of unrecyclable plastics rather than coal, avoiding carbon dioxide emissions and impacts from coal mining.

Even trash that is put in landfills can provide some value. Cities can collect methane that rises as the waste decomposes, which is an obvious improvement over flaring (burning off) the gas or simply letting the methane waft up into the atmosphere, where it traps much more heat than the equivalent amount of carbon dioxide. Power generators can convert the collected gas into electricity. In British Columbia, Vancouver's landfills extract the methane and burn it to heat nearby greenhouses that grow tomatoes.

Even then, landfills are still leaky. That leakiness inspired Vancouver, which has pledged to become the greenest city on Earth, to give residents separate bins for trash and organic matter (food scraps, yard clippings and tree trimmings). Officials expect citizens to use them properly and deploy city inspectors to check that waste haulers are dumping refuse that is separated correctly. The city produces methane from the organic waste while generating solids known as amendments that can make soil more fertile. These solutions solve multiple problems at once – saving money for energy that would otherwise have been purchased, reducing the need for expensive landfilling and avoiding unnecessary use and damage of land – while improving agriculture.

Austin does something similar with its wastewater sludge, passing it through anaerobic digesters to make biogas it sells or uses onsite for generating heat. It converts the remaining solids into a popular soil amendment known as Dillo Dirt (a reference to the armadillo, one of the local creatures). The city earns money by selling the Dillo Dirt, offsetting some of the cost of treating wastewater. Although composting is a growing and popular trend among residents – and one worth pursuing – doing it poorly can actually lead to more methane emissions. For Austin residents, it often makes more sense for residents to put food scraps down the drain and through a grinder so that the city's industrial-scale harvesters at the wastewater plant can do the work of the composter but with greater efficiency.

Waste heat is another big opportunity. Harvesting it is difficult because low temperatures are hard to convert into electricity. NASA developed thermoelectric generators to do this on its spacecraft, but the technology is expensive and inefficient. Nevertheless, advanced materials that can more effectively convert heat to electricity are coming. A place to start is the hot wastewater that goes down the drain when we wash our clothes, dishes or bodies. Sandvika, a suburb of Oslo, has massive heat exchangers along city waste pipes that extract heat to warm dozens of nearby buildings or defrost sidewalks and roadways. By turning on heat pumps in the summer, it can use some of the heat to cool those same buildings. Vancouver liked the idea so much that it repeated the concept, using wastewater to heat hundreds of buildings and the Olympic Village.

Taking that idea further is the Kalundborg Symbiosis in Denmark, a leading example of closed loop thinking. The industrial park has seven companies plus municipal facilities – centered on electric, water, wastewater and solid waste facilities – that are interconnected such that the waste from one is an input for another. Pipes, wires and ducts move steam, gas, electricity, water and wastes back and forth to improve overall efficiency and reduce total wastes, including CO_2 emissions. For example, wastewater from the oil refinery flows to the power plant, where it is used to clean and stabilize fly ash from coal combustion. The refinery also sends waste steam to Novo Nordisk, which puts the heat to work for growing about half the world's supply of insulin with bacteria and yeast. The entire industrial park looks like a living, industrial organism. And it has demonstrated economic growth with flatlined or reduced emissions.

Data-Driven Decisions

Can the Kalundborg symbiosis model be replicated on a larger scale, for cities worldwide? Yes, but only if we make cities smart. An industrial park is flexible because it has only a few tenants and decision-makers, but a city has many individuals and organizations making independent decisions about energy, water and waste every day. Integrating them requires a cultural shift toward cooperation, boosted by advances in smart technologies. "Smart cities" will depend on ubiquitous sensing and cheap computing, compounded by machine learning and artificial intelligence. This combination can identify inefficiencies and optimize operations, reducing wastes and costs while operating all kinds of equipment automatically.

Thankfully, making cities smart is an alluring objective for planners who want to accommodate higher densities of people without diminishing quality of life. For example, in 2015 in India, where population and public health problems are severe, Prime Minister Narendra Modi announced his intention to convert 100 small- and medium-sized municipalities into smart cities as a possible solution.

The "smart" moniker itself is an accusation that most cities are dumb. That accusation sticks because municipalities rife with waste seem to be operating blind. In 2015, the U.S. National Science Foundation launched a major research initiative called Smart & Connected Communities to help cities make better use of data. That name, by the way, indicates that intelligence is not enough: Interconnections among systems and people matter too.

Smart cities rely heavily on big data gathered from widespread sensor networks and advanced algorithms to quickly gain insights, draw conclusions and make decisions on those data. Connected networks then communicate those analyses to equipment all across the city. Smart meters for closely tracking electricity, natural gas and water use by time of day, household and industrial appliance are an obvious place to start. Realtime traffic sensors, air-quality monitors and leak detectors are also at hand. The Pecan Street consortium in Austin is collecting data from hundreds of homes to learn how access to such data streams might help consumers change their behaviors in ways that reduce consumption while saving costs. Cities such as Phoenix and military bases such as Fort Carson in Colorado have pledged to become self-sufficient users of energy and water and net-zero producers of waste. Achieving those ambitious goals will require a lot of interconnected data.

Better transportation may give urbanites their first glimpse of a smart city's benefits by cutting wasted time. Reducing the footprint of transportation means cleaning up the fuels, making the vehicles more efficient, reducing trip distances and duration, increasing vehicle occupancy and cutting back on the number of trips. If people live close to their work, they can walk or bike or use mass transit. Studies show that building protected bicycle lanes leads to dramatically increased ridership, and because bicycles require so little space, compared with cars, they can reduce congestion on the roads.

A driverless city will also free up wasted space and time associated with parking. With shared or autonomous cars in constant motion instead of private cars that are parked at home and work, the number of parking spaces can be restricted dramatically, opening up wasted space and easing congestion further. Researchers at the Center for Transportation Research at UT-Austin used sophisticated models to determine that shared, autonomous vehicles would lessen the number of cars needed in a city by an order of magnitude and would cut emissions, despite causing a slight increase in total miles traveled because the vehicles would stay in motion. Instead of wasting their time driving, commuters can rest, read emails, place phone calls or conduct other business. That work can create economic value – and trim a person's office hours so he or she can get home earlier for dinner.

Making our infrastructure smarter is certainly the key to solving basic problems such as leaky water pipes. Identifying leaks should be easy if meters are distributed throughout a water system to track flows and readily pinpoint the amount and location of those leaks. Researchers in Birmingham, England, developed a system with tiny pressure sensors that use a small amount of power to frequently check for and detect leaks in water networks, a big improvement over the old technique of waiting for someone to call and complain that water is shooting like a geyser out of the road. Someday we might even send smart robots down the pipes to repair the problems.

High-performance sensors will also let us find and predict natural gas leaks before accidents happen. Gas leaks are not only bad for the environment and a waste of resources but dangerous, as we see in headline-grabbing explosions in urban areas with aging infrastructure.

It is hard to know where smart, waste-conscious cities may arise. I imagine a likely candidate will be a Midwestern town with a million people or more that needs to reinvent itself because its economy was gutted decades ago. Indianapolis comes to mind, in part because it needs to rebuild water, wastewater and sewer systems based on bad decisions a century ago. The city has been investing in its downtown and is on the rise. Pittsburgh is leveraging its existing assets – a vibrant urban core, city pride, forward-looking leadership from its mayor, the strength of Carnegie Mellon University and other hotbeds of innovation – to go from being defined by its smokestacks to being defined by its brainpower. Indeed, Uber launched its autonomous vehicle service there. Columbus, Ohio, which is the state capital and home to a major university, is another place to look for cutting-edge experiments in becoming smart. In 2016, the U.S. Department of Transportation awarded Columbus a $40million grant to reinvent its approach to mobility.

Getting From Here to There

Turning profligate cities into places that reduce waste and reuse what is left will not be easy. Integrated R&D investments from the federal government have to be combined with practical policies from all levels of government.

Investment has to be socially savvy as well. Studies show that R&D for smart cities has focused more on technology than what the citizenry needs. Done the wrong way, the benefits of a smart city might accrue to those who already have Internet connectivity and access to advanced technologies, which would only widen the technology gap on top of other socioeconomic divides.

Municipalities also need to help residents become smarter citizens because each individual makes resource decisions every time he or she buys a product or flips a switch. Access to education and data will be paramount. Connecting those citizens also requires collaboration and neighborly interactions through places like parks, playgrounds, shared spaces, schools, and religious and community centers—all of which were central tenets of centuries-old designs for thriving cities. The more modern and smart our cities become, the more we might need these old-world elements to keep us together.

Wasting an Opportunity

MECHANICAL ENGINEERING, NOVEMBER 2014

> *Americans generate a lot of trash and stick most of it in landfills. We need to recover more value from what we throw out.*

THE WORLD OF WASTE IS INNOVATIVE IN WAYS MOST AMERICANS would never imagine. Advanced recycling centers in the United States, for instance, use multimillion-dollar machines with high-tech conveyor systems to automatically sort different materials using lasers, optical scanners, blowers, magnets, rakes, crushers, vacuums and other devices. Unsorted trash enters the system, and neatly separated bales of like materials exit.

Nonetheless, in spite of generating more than 20 percent of the world's trash, Americans are pretty unsophisticated in the way we deal with it. Europeans, conversely, produce less trash per person and send very little of what they do make to landfills. Why? To begin with, they do not have much space for landfills. Europeans also have a different cultural approach to resource management, aiming to use less and make each ounce of material go further. Americans often associate consumption with power and conservation with weakness; in Europe, it's the other way around.

The European Union has strict regulations about the end-of-life for different consumer goods, including electronics and automobiles. These regulations mostly prohibit landfills as an option for a majority of the materials and put a responsibility on manufacturers to create designs that can be disassembled and separated into useful materials after the product life. The result? Many of those expensive, high-value systems that automate the recycling process are imported from Europe.

Compared to Europeans, who build high-tech machinery to recover valuable materials, Americans seem backward. We're essentially burying money in the ground. It's time we realized it.

Many plastics can be pelletized into solid fuels that displace coal or liquefied into fuels that displace petroleum products. Fibers can be used as input for building materials. Metals can be melted down and used again.[11]

Cities still pay recyclers millions of dollars to take away trash. But those costs are dropping as the separation and recycling technologies improve and the value of the harvest materials increases.

Changing the way we look at trash will require changes at every level of government. At the state and local levels, decision-makers must recognize the value in trash; we could be charging companies for the right to "mine" our waste streams instead of paying them to haul it away. Federal R&D can bring down the costs of sorting and recycling our trash

11 See "Trash to Treasure" on p. 47, for example

and help improve scrubbers for trash incinerators. Federal policymakers can also implement strict, European-style end-of-life requirements for durable goods.

The U.S. military is already operating in the new waste paradigm. For the Department of Defense, waste is a strategic liability; in some forward operating bases, every pound of waste needs to be trucked off site. And those trucks are soft targets that put human lives at risk, so reducing trash isn't just good for the environment, it also provides operational security.

Stateside, Fort Bliss (Texas) and Fort Carson (Colorado) are both pursuing zero-waste initiatives. Doing so reduces the footprint of those bases and helps the military develop the expertise it needs in theater.

If reducing our waste and managing it better is good enough for our soldiers in harm's way, it's good enough for civilians here at home too.

The Carbon Dioxide We Dump Into the Sky Is Just Another Kind of Garbage

SCIENTIFIC AMERICAN, DECEMBER 2019

> *We should clean the sky up as we do ordinary trash – and it wouldn't cost much more.*

FOR DECADES IN THE CLIMATE CHANGE DEBATES, WE'VE HEARD that reducing carbon emissions will cost society too much money. Indeed, curbing CO_2 emissions is expensive. For a large economy like the U.S., which emits about 5 billion tons annually, and a cost of $10 to $100 per ton for removing CO_2 from the atmosphere, the price tag can range anywhere from a staggering $50 to $500 billion. Every year.

That sounds like a lot, right?

Not so fast. What if we look to other parts of society where we already invest in efforts to clean up after ourselves? It turns out that in other waste management areas – trash disposal and wastewater treatment – Americans are already investing large sums of money. And the payoff isn't just cleaner cities and streams. Our commitment to cleaning up after ourselves creates hundreds of thousands of jobs and yields long-lasting benefits to our economy and public health.

It's time we applied the same mindset to our airborne waste streams.

For inspiration, consider garbage. Each year the U.S. generates more than 290 million tons of municipal solid waste. If we didn't dispose of it properly, the American landscape would be littered with smelly dumps contributing to dysentery, diarrhea and respiratory disease. That's the main reason trash collection and disposal is a standard requirement for American homes, businesses and neighborhoods. All told, we readily agree to spend about $200 billion a year on solid waste management.

Our investments in wastewater pay similar dividends. Years ago, Americans pumped untreated sewage and industrial pollution directly into waterways, creating all sorts of problems. In response, cities and the federal government passed legislation reducing pollution and requiring treatment and cleanup. We now spend more than $100 billion each year for our water and wastewater treatment systems – a bargain when one considers the nightmare of societies ravaged by waterborne diseases.

If we're willing to pay to clean up our solid waste and liquid waste, why not our gaseous waste? Instead of waste collection via pipes and trucks, we would have CO_2 collection via scrubbers and direct air capture.

How would we do this? First, we should make the problem simpler by cutting our CO_2 pollution. The organization Energy Innovation estimates U.S. emissions could be halved by 2050 by using policies that have been demonstrated in other countries – approaches such as a carbon tax, clean electricity standards and better efficiency standards.

That would still leave a lot of CO_2 to clean up – about 2.5 gigatons per year at present emission rates.

Thankfully, CO_2 scrubbing technologies are improving rapidly. Recent analysis indicates that direct air capture of CO_2 would cost about

$35-75/ton in the next 20-30 years. Compared to the $700/ton we spend managing our solid waste, that's a bargain.

At those rates, scrubbing all the excess CO_2 emissions from our common skies would require $90-190 billion per year. That's right in line with what we invest now for our other waste streams.

Much of this excess CO_2 would be buried underground, but there are geological limits. Fortunately, here too, the way we handle garbage holds suggestions for CO_2. Just as we reduce garbage landfills by recycling materials, we could potentially recycle some of our excess CO_2, converting it into useful products such as chemicals, fuels and building materials.

Americans didn't commit to cleaner landfills, streets and streams overnight. The consensus for cleaning up developed slowly as the perils of pollution became painfully clear through environmental catastrophes and public health crises.

One thing that experience taught us is that the price of cleaning up after ourselves isn't too high. Implemented over time, at the federal and local level, it's manageable – and well worth the investment.

Ultimately, societies that deal with their waste are wealthy and healthy. Countries that don't deal with their waste, letting it accumulate and pollute their environment, are sick and poor. Personally, I prefer for the U.S. to be wealthy and healthy, and that means we should deal with our waste. All of it.

The Solution to America's Energy Waste Problem

FORTUNE, DECEMBER 2017

By Joshua D. Rhodes and Michael E. Webber

> *The U.S. economy consumes a tremendous amount of energy. Energy is a good thing – critical to our prosperity, wealth, health and stability. The problem is we waste too much of it.*

SIMPLY PUT, THE U.S. ECONOMY IS NOT AN EFFICIENT ECONOMY. Incredible as it may sound, we waste about two-thirds of the roughly 100 quads (quadrillion Btu) of energy we consume each year. Most of this waste is due to the burning of fuels. When we burn fuels, whether it's in a car's engine or our home oven, most of that energy becomes useless waste heat.

Our economy almost always wants more energy, but there are limits to some of our energy resources. The best way to make those resources go further is to first solve our energy waste problems.

So what needs to be done? The answer is simple: Improve the energy efficiency of the economy as a whole. And the best way to improve the economy's efficiency is through deep electrification, or taking some of our standard activities that are powered by burning fuels and using electricity instead. In the old days, that meant replacing kerosene lamps with electric lightbulbs, but today that means replacing gasoline-powered automobiles with electric vehicles and industrial boilers that burn natural gas to make steam with more advanced devices.

We get our energy from primary sources such as uranium, water, wind, solar radiation, coal, natural gas, oil, biomass and the ocean's tides. Some of these fuels are used directly, such as natural gas for cooking. Others are first converted into another form so that they are more useful: Oil is refined into petroleum products such as gasoline for cars, and coal or natural gas are burned to generate electricity.

This last form of energy – electricity – is particularly valuable because it is so versatile; it provides clean motion, heat, light and information. It is an essential and pervasive presence throughout the developed world and a highly desired commodity throughout the developing world. Electrification is also a critical avenue to address climate change because it enables the integration of low- or zero-carbon energy sources such as wind, solar, nuclear fuels, natural gas and geothermal heat.

But an electric economy is also an energy-efficient economy, producing less waste, lower costs and less pollution.

For instance, electrification of transportation would be transformative. The transportation industry consumes the most energy of all end-use sectors, about 29 percent of total U.S. energy consumption, mostly in the form of petroleum products burned in internal combustion engines operating with about 25 percent efficiency. This means that of the 20 gallons you put in your gas tank, only five are used for propulsion, and the other 15 are wasted. While driving, your engine heats up and ejects heat into the atmosphere – wasted energy that does not transport you.

According to our analysis, if all 3 trillion miles that our cars and light-duty trucks traveled in 2017 were instead provided by electric vehicles that are 70 percent efficient, we would reduce our economy's rejected, or wasted, energy from 66 quads to 59 quads. That's the equivalent of improving the entire economy's efficiency from 32 to 34 percent. That might seem insignificant, but a small change in a big number can make a huge difference.

Electric propulsion is much more efficient than heat-based propulsion – only 4.8 quads of electricity are required to displace 16 quads of petroleum to move our vehicles, for instance. If that new or extra electricity demand were met by modern natural gas combined-cycle power plants with about 55 percent efficiency, we would need to burn only another 8.8 quads of natural gas to displace the 16 quads of petroleum, a primary energy reduction of about 9 percent.

To take this scenario one step further, if this added electricity were instead produced by zero-emission technologies such as nuclear, wind or solar, it would also reduce carbon dioxide emissions more than 1 billion metric tons (out of nearly 5 billion tons of annual emissions). If the added electricity were to come from old, inefficient coal plants, overall carbon dioxide emissions would increase by 166 million metric tons.

Electrifying other uses of energy beyond transportation – such as heating, cooking and industrial activity in our homes, businesses and factories – also offers many benefits. Steel mills that use electric arc heating are more efficient than old-fashioned ones that burn coal. Oil and gas operations can use down-hole electric submersible pumps to increase productivity and avoid pneumatic controllers that vent greenhouse gases into the atmosphere.

As the political debate rages about how to reduce pollution while keeping energy affordable and reliable, energy efficiency through electrification has broad appeal because it saves money and reduces environmental damage. Though it won't solve all of our energy needs – especially for sectors like chemicals, industrial heat, aviation or marine shipping – deep electrification is a direct pathway to an efficient and more productive economy.

Section II

Safety Needs

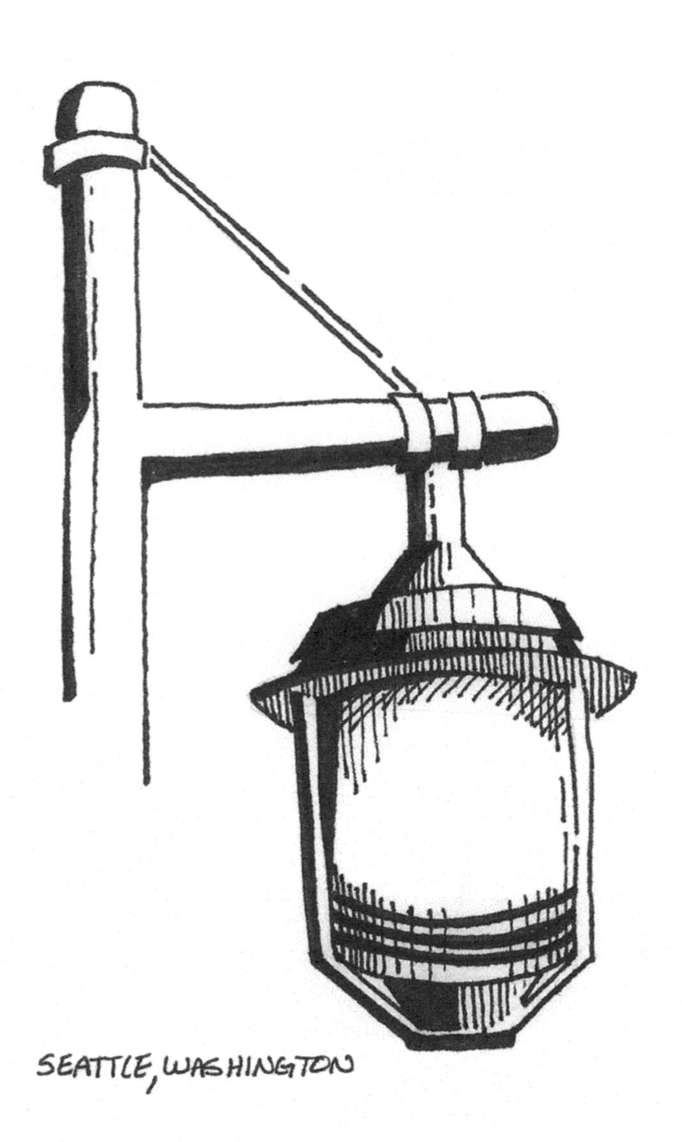
SEATTLE, WASHINGTON

Chapter 3

Safety Needs I: Energy & Education

We Need More Than STEM

MECHANICAL ENGINEERING, JANUARY 2020

> *The energy sector's challenges require education in the arts as well as science, math and technology.*

THE UNITED STATES IS TECHNOLOGICALLY ADVANCED: WE PUT humans on the moon and invented the Internet. Yet surprisingly large numbers of Americans don't understand such basic science concepts as what causes the seasons or whether the Earth orbits the sun. This cluelessness would be just a national eccentricity if the U.S. didn't have decisions to make on complex energy policy issues, such as the role of nuclear power or the transition away from carbon fuels.

What's more, the energy industry needs a pool of workers skilled in science, technology, engineering and mathematics (or STEM) to navigate that impending energy transition. While no one is sure what the exact demand will be – perhaps fewer chemical engineers to refine petroleum and more mechanical engineers to design microturbines, or electrochemists to develop better batteries – our energy transition will surely benefit from STEM-capable people.

Those two factors, plus the need for a better-informed electorate and the demand for a more technology-savvy workforce, have led policymakers to present STEM education as the go-to panacea for many of our educational and sustainability woes. Across the country, university humanities programs are facing budget cuts, and many thought leaders sneer at the idea of liberal arts.

This is a mistake.

I'm a STEM guy through-and-through. My father is a chemistry professor, and I have degrees in aerospace, electrical and mechanical engineering. I now have a courtesy faculty appointment in the University of Texas Austin department of Civil, Architectural and Environmental Engineering. The classes I teach are cross-listed in chemical engineering and geosciences. I'm steeped in STEM, love STEM topics, and think STEM students are remarkable.

But I do not think STEM is enough.

STEM gives students the formulas, principles and theorems needed to turn raw information into a suitable answer, all while operating within the constraints of the physical universe. It's a remarkable capability, and we should be happy to have their brains deployed for society's greatest challenges. But we need more: We need the arts, both the fine arts and the liberal arts.

STEM teaches us analytical thinking – how to solve a problem. But the liberal arts teach us critical thinking, or why we should solve the problem; and the fine arts teach innovative thinking, inspiring us to take action and providing the creativity we need to come up with new approaches.

For example, present engineers with a challenge and their solution is almost always that we should build something new. Need to control floods or provide more drinking water? Build dams. Is energy scarce? Build power plants. There's nothing wrong with this tendency: I'm an engineer, and I see it in myself.

Sometimes, though, we would be better served if we stepped back and reevaluated the problem. Faced with a problem like endemic flooding, critical thinkers would assess whether we needed new construction at all, or whether we might use urban policy and architecture to protect our floodplains and waterways to minimize flooding and store water. And creative thinkers might reframe the question entirely and create informational campaigns to educate the public about the powerful role of conservation to reduce energy or water demand.

We will need all of those skillsets to master the multiple global challenges of this century. Ultimately, everyone has a role to play. Tackling climate change and managing the energy transition in an elegant way will take all hands on deck. It also needs new kinds of thinking. That means we shouldn't cast aside any disciplines or people who can be part of the solution. And we need to infuse whatever solutions we develop with a deep understanding of the human condition.

I mentioned that I'm from a STEM family, but my sister is an art teacher. She likes to say, "The EARTH without ART is just 'eh.'" Indeed, art – both the liberal and fine arts – makes the world more meaningful and interesting. Taking a cue from her, we should go from STEM to STEAM (the extra letter is for "art") to give us the critical capabilities we will need.

Better Tools for Energy Literacy

MECHANICAL ENGINEERING, APRIL 2014

> *Improving energy literacy should be one of the most critical priorities for the United States.*

THE OLD METHODS OF TEACHING – USING CHALK AND TALK – aren't giving citizens a deep enough understanding of complex energy issues.

Just about every sector of society and the economy is affected by energy-related policies. But because scientific fundamentals as well as economics, politics, law and culture underlie what energy can – or can't – do, the general public has a hard time engaging in the policy debates in a meaningful way. Too often, the public debate gets reduced to bumper sticker slogans such as "Drill, baby, drill," or "Don't frack the planet."

Improving energy literacy should be one of the most critical priorities for the United States. As an energy educator, I believe we can use innovation to bridge the gap by supplementing traditional classroom instruction with engaging and dynamic outreach initiatives. In 2013, I had the opportunity to practice what I preach by experimenting with new teaching technologies and methods across a variety of different media.

One way of doing this was via television. I hosted "Energy at the Movies," a nationally syndicated PBS television special that featured clips from different movies to teach the history and science of energy. Movies such as *Silkwood, Syriana* and *Promised Land* contain useful imagery and scenes related to energy topics, and my role as the educator was to point out where Hollywood got the science right – or wrong. Early feedback on the special was so positive, with telecasts reaching hundreds of thousands of viewers, a follow-on documentary series was developed.[12]

One of the benefits of working with public television is the ability to reach a large audience of general interest viewers and prospective minority STEM students. Expanding this approach could help address the needs of underrepresented groups.

---

12 Season One of the documentary series "Power Trip: The Story of Energy" aired six episodes on PBS starting Earth Day 2020 and has subsequently been broadcast more than 7,000 times with additional distribution on Amazon Prime, Apple TV, and on networks in dozens of countries, reaching millions of viewers. Season Two, with six additional episodes, was released starting September 1, 2023. For more information, visit www.powertripshow.com

I also taught a successful massive open online course, or MOOC, in collaboration with EdX in fall 2013 titled Energy 101. More than 44,000 students from 173 countries enrolled. Nearly 5,000 students completed the course, resulting in a 13 percent completion rate, which was more than twice the rate for a typical MOOC.

During my first virtual "office hours," I received questions from students on every continent except Antarctica. In all, this MOOC expanded global reach for energy literacy and STEM education. The platform brought a college course to people who would not easily have access otherwise. In fact, a few high schools used it. These two experiments – the national television show and the global MOOC – offer a few lessons learned.

First, support and financial backing from the school – from the department chairs up to the university president – is needed before educators can experiment with multimedia teaching tools. Fortunately for me, the University of Texas at Austin and the Cockrell School of Engineering are both very supportive.

Creating educational content for the general population or for a global audience makes for better educators. Energy 101 is based on a graduate course retooled every few semesters at UT-Austin. However, video is forever, so close content scrutiny and a highly coherent lesson plan were vital for the MOOC's success. Also, because many MOOC students are outside the United States, teaching a MOOC requires an international lens. All this focused my teaching and forced me to reframe some of the arcane details of American energy policy in a way that could be accessible to everyone.

Overall, the MOOC was an effective experiment with energy literacy curriculum. However, until online assessment capabilities improve,[13] the traditional classroom will continue to serve a critical

---

13 Since 2013, online educational technologies *have* improved significantly, opening up the possibility for a hybrid classroom format (some students in person, some students online) as a common configuration.

function for not only distributing critical information but for assessing students' learning objectives.

Indeed, MOOCs are probably misnamed. They function less as courses and more as open textbooks – a massive open online textbook, if you will. Turning these online materials into high-quality, interactive textbooks is the right direction for the future.[14] Making these dynamic teaching materials available to a new, global audience hungry for content is a powerful way to expand STEM education, and it can enhance energy literacy. Both of those outcomes would benefit society.

It's Time to Shine the Spotlight on Energy Education

THE CHRONICLE OF HIGHER EDUCATION, JANUARY 2012

By Michael E. Webber and Sheril R. Kirshenbaum

> *Isn't it time for this country's colleges to take significant steps toward developing a new approach to energy education?*

ADVANCES IN THE ENERGY SECTOR THESE DAYS MUST CLEAR multiple hurdles; in particular, they must meet global needs without compromising national security, degrading the environment or impinging on the economy. But unfortunately, innovation in the large-scale energy field still seems to plod along through the invention of technologies designed to achieve singular purposes. Isn't it time for this country's colleges to take significant steps toward developing a new approach to energy education?

14 Following my own advice, in late 2014, we converted Energy101 into an interactive online textbook (www.energy101.com) that is updated annually. It is used as a reference for continuing education or traditional instruction on multiple continents by industry, higher education and high schools.

In 2012, for the most part, higher education for students interested in energy lacks the cross-disciplinary curriculum that they critically need. So, we propose the adoption of energy departments on college campuses, departments that would tie seemingly disconnected fields of the sector together. An army of renaissance women and men would emerge who would over time make up a cutting-edge labor force able to understand and articulate not just the science and business of energy, but the political, technical and social issues involved in finding solutions to our energy needs.

Across the country, undergraduates are being ushered through an outdated and compartmentalized system in which the education has not kept up with scientific advances. Energy is poorly defined at institutions of higher education, appearing to be an ambiguous professional pursuit or a subset of umbrella departments such as petroleum engineering or geosciences, which tackle only a single slice of the energy pie. Students must typically choose to enroll in a single department where they are exposed to narrow perspectives of the energy sector and do not obtain a comprehensive understanding of what lies ahead.

For example, engineering departments offer courses in technical problem solving; political science departments focus on the theory of governance; policy students learn to craft legislation; economics students analyze what motivates personal energy choices; and communication majors acquire the skills to translate complex language. At the same time, biologists discuss environmental impacts, and sociologists develop the tools to influence behavior. And so on. But each discipline acts as if its work occurs in isolation from other fields. Thus, in the process of earning a single specialized degree, students are limited in their exposure to other related fields. Under this paradigm, graduates do not leave with a comprehensive understanding of energy, and this traditional model slows progress in an increasingly globalized world.

Consider the typical petroleum engineer: Their career will be deeply affected by policy and economics, yet we do not train them to participate in the process governing their actions. Meanwhile, policy makers and economists are generally educated about drilling by lobbyists and special interest groups in isolation. Thus, what we have is a broken system that limits meaningful contact among those with the greatest expertise in each area.

But there is some good news on the horizon. Over the past decade, many top tier universities, including our own (the University of Texas Austin), have been moving toward more interdisciplinary certificate programs, specializations and degrees where the arts and sciences converge. These include a certificate program for graduate students called GPPES (Graduate Portfolio Program in Energy Studies) and a minor in sustainability for undergraduate engineering students.

[As of the writing of this essay in 2012,] MIT, Duke University, the University of British Columbia and UT-Austin have developed courses or degrees in response to the changing energy landscape. Some are focused on energy, some are multidisciplinary, some are graduate, and some are undergraduate. It's clearly time to build on those efforts and promote the development of robust academic departments of energy that would enable students to engage with a vast array of related topics, from sustainable agriculture to international security. [In the decade since this piece was written, more universities have created new multidisciplinary programs in energy or sustainability, converted petroleum engineering programs to energy engineering and enabled more offerings along the lines of what we recommended here.]

Such departments would bring professors together from a variety of disciplines across campus to develop an organized energy curriculum. No longer would economics and engineering advances be isolated from each other. They would be hubs for a variety of areas tackling theoretical and practical questions that involve subjects like consumption, smart grids, renewables, waste and more.

Energy departments would highlight history and the complex challenges ahead, while emphasizing social, economic and political environments domestically and abroad. Graduates entering the work force with such interdisciplinary skills would be ideally suited for leadership positions with strategic responsibilities, armed with a firm grasp of the challenges ahead and the experience to act as truly global citizens.

New Engineering Thinking for a New Climate

MECHANICAL ENGINEERING, JUNE 2016

Reducing carbon emissions is the engineering challenge of the 21st century. To meet it, mechanical engineers will have to find new approaches to familiar problems.

EVERYTHING WE THOUGHT WE KNEW ABOUT MECHANICAL engineering has changed, thanks to the agreement that capped the 2015 United Nations Climate Change Conference in Paris. While the agreement is nonbinding, the direction is clear: "A balance between anthropogenic emissions by sources and removals by sinks of greenhouse gases in the second half of this century."

Simply put, we need to drive down net emissions of carbon before 2050.

The Paris Agreement has, I believe, become one of the primary geopolitical contexts of the 21st century. The next generation of professionals, including engineers, scientists and policymakers, will be affected by the call to sharply reduce carbon emissions. And those reductions have to start with the power sector, which is responsible for a quarter of global carbon emissions and more than 30 percent of those in the United States.[15]

Engineers will have to rethink and reengineer the power system from top to bottom. And to accompany this new approach to reengineering, we must transform engineering education as well. Because climate change has put carbon emissions at the center of the global political discussion, we need to produce engineers who not only discover new solutions to enduring problems, but who can talk to multiple stakeholders from wide-ranging backgrounds.

15 Since this piece was written, power sector emissions in the United States have dropped such that transportation sector emissions now exceed them.

That said, the rethinking isn't something that only affects the next generation of mechanical engineers. It starts with all of us, and it starts now.

The Next Transition

One place where teaching–and thinking–must change is how engineers grapple with sweeping, systemic changes, especially in energy technology. Too often, when we look to the past, we constrain ourselves to searching for specific solutions that we can reapply. Instead, we need to think about large-scale transitions, how they happened and, most importantly, why.

The widespread, global decarbonization of the power sector called for by the Paris Agreement won't be the first time society has switched from one energy system to another. Over the last 200 years, we've transitioned from wood to coal to petroleum. Each transition was an improvement since each new fuel source was higher performing and less dirty and carbon intensive than the one it supplanted.

Today we are on the cusp of another transition, as natural gas is poised to surpass petroleum as the dominant fuel in U.S energy consumption. And that follows the decarbonizing trend of the earlier transitions, since combusting methane releases less carbon than does burning coal.

The lessons from previous transitions are worth keeping in mind.

To start with, shifts from one dominant fuel to another have taken decades, so if we are going to meet the goal set for the second half of the century, we have to start today. It isn't just new power plants that must be built, but also an expansive logistical infrastructure that moves fuel and electricity from producer to consumer.

Next, the earlier transitions occurred because of performance – for example, coal is easier to handle and more energy dense than wood. Today, natural gas use eliminates the fuel-handling and preparation requirements of a typical pulverized coal plant. As we develop low-carbon solutions, engineers have to make sure that new technologies have performance advantages, otherwise it will be hard to convince industries to switch.

Today, for instance, utilities looking to add generating capacity don't need to have their arms twisted to opt for natural gas combined-cycle power plants over legacy coal-fired thermal plants; such plants are twice as efficient, have about half the operations and maintenance costs, and can be built for just a fraction of the cost of a new coal facility. So many gas turbines (both simple and combined-cycle) have been added to the U.S. grid in the preceding 10 years that in 2015, power plants burning cheap natural gas were responsible for as much electricity generation as traditional coal plants. As recently as 2006, gas generated only half as much as coal.

And as we promote the next transition, we can't lose sight of the fact that today's solutions can become tomorrow's problems. The engineers who helped bring coal and oil to market actually solved some long-standing ecological problems related to deforestation and whaling. They may never have dreamed that their new technology would one day be considered "dirty." Engineers need to examine the new energy solutions we come up with to weed out any latent complications that may grow to become societal problems.

To see how those lessons come together, consider the possibility of a complete transition to wind and solar. By some measures, those renewables offer significant performance advantages over the conventional options – the wind and sun are essentially free and inexhaustible, and they obviate the need for fuel handling entirely. Plant operators won't need to worry about rising fuel prices or fret about fuel supply contracts. And contrary to long-standing concerns about the ability for renewables to scale up, the last few years have seen wind and solar power surging onto the U.S. grid, going from 55,000 gigawatt-hours in 2008 to 240,000 in 2015, according to the U.S. Energy Information Administration.

At the same time, wind and solar power still present challenges that have yet to be resolved. Each vary both seasonally and from minute to minute based on the weather and astronomical conditions. Those dynamic attributes create a lot of stress on the grid as other generating plants get used less (increasing their costs), yet are called on short notice to back up any changes in wind and solar output. While wind and solar power produce electricity, which is what utilities want, the production isn't tied to actual demand; for instance, a lot of wind power is produced in the middle of the night when no one is awake to use it. And even at moderate scales of deployment, renewables have had some unexpected environmental impacts – such as bird strikes – that have soured some on their expanded use. Those are all solvable problems.

Multiple Choices

The challenge of the Paris Agreement differs from earlier energy transitions in an important way: This shift is being intentionally accelerated, rather than occurring accidentally at a slow pace as before. But we have to teach engineering students that responding to policy pressure isn't a new thing for the power sector.

A generation ago in the United States, for instance, the formation of acid rain from sulfur and nitrogen oxide emissions became an intolerable environmental problem and prompted a public policy solution. Regional limits were placed on NOx and SOx power plant emissions, and operators of noncompliant power plants had a few choices: They could retrofit, switch fuels, retire or pay a fee for credits from cleaner power plants.

Operators of some power plants opted to install large-scale, very expensive scrubbers to remove the pollutants from flue gases. That approach kicked off a wave of innovation and gave engineering firms in the United States a competitive advantage, as environmental scrubbers became a multibillion-dollar export business. Other operators found it easier to switch fuels from Eastern, high-sulfur coal to low-sulfur coal from the West. Today, Wyoming produces more coal than Kentucky and West Virginia combined.

But operators of some old, creaky, inefficient coal-fired power plants couldn't justify either of those moves. Instead, they shut down their plants and built new more-efficient natural gas power plants in their place.

The multiple-prong approach worked. The acid rain problem, while not completely solved, has been dramatically improved. According to a 2004 assessment by the U.S. Environmental Protection Agency, dealing with acid rain took less time and money than anticipated – and provided $40 billion of health benefits for every $1 billion of investment in scrubbers and power plants. Along the way, grid reliability improved and electricity prices stayed level.

The success story of acid rain regulations should focus our thinking when we approach the much larger challenge of post-Paris decarbonization. We must look hard at the existing technologies and energy sectors and apply our new thinking to the question of how much we want to retrofit and adapt by installing new scrubbers, and how much needs to be built from scratch so we can switch fuels entirely.

For example, even with the mandate of decarbonization, there could still be a place for coal. That fuel has considerable advantages, such as abundant domestic resources, predictable pricing and ease of storage in large piles on-site at power plants. But for coal to maintain its position, we will need to clean it up with modern designs and carbon capture systems.

If plant operators can't make new coal plants pencil out, they could turn to another familiar technology: nuclear power. Massive, cheap and reliable nuclear power plants have ably served baseload demand for decades, and new plants are springing up in Asia. But questions regarding cost overruns on new construction, public safety, waste management and weapons proliferation still haven't been fully resolved. Also, since traditional nuclear power plant designs are heavily dependent on water for cooling, they are vulnerable to reduced performance as increasingly frequent and intense droughts and heat waves strain the availability of cool water.

To make nuclear power a bit more embraceable, we must work to develop advanced fuel cycles, small modular reactors, dry cooling and passively safe designs.

And it will make sense to integrate nuclear power with water production systems so that waste heat is used for water treatment or desalination.

A bigger challenge is to find ways to enable nuclear power plants to operate more flexibly as renewables become an ever-larger part of the energy mix. Engineers will have to optimize designs capable of rapid ramping and, critically, find ways to incorporate novel storage approaches. Thermal energy storage with molten salts, for instance, might be more appropriate for nuclear power, though many materials and heat transfer challenges remain.

But much like the plant operators of the 1990s, we may have to face retiring and replacing large portions of the power sector. Fortunately, we have many viable alternatives. Clean and lean natural gas combined-cycle power plants are ready today. And wind and solar are growing in popularity even if they are not – yet – a complete solution.

Intermittency and Beyond

Handling the influx of power from intermittent sources such as wind and solar is going to require mechanical engineers to rethink the transmission and distribution system. The amount of capital assets tied up in the wires and poles of the transmission and distribution (T&D) system – about $4 trillion worldwide – approaches the $6 trillion capital value in power plants, so building out a bigger, better, smarter grid is going to be a major challenge of its own.

Engineers will have to think differently about T&D to make this new grid work. Today's conventional approach to the grid uses a load-following mentality. That is, we turn our lights and appliances on and off all day long, which changes the load on the grid. Power plants are then dispatched up and down to follow the load and to balance the supply and demand. But that approach makes it difficult for operators to handle power from nondispatchable wind farms and solar plants.

Rather than expecting power plants to operate when we need them, we could instead schedule our demand to match the forecast supply of wind and solar. There are many operations – water heating, pool pumping, water treatment, noncritical data centers, and so forth – that can be performed flexibly over the course of the day. Connecting machines and appliances to the Internet via smart meters could enable grid operators to remotely dial down demand as necessary and provide valuable reliability services.

But tomorrow's grid issues go beyond intermittency. Adding renewables on a large scale requires long-distance lines connecting cities to distant plains and deserts as well as new technology and innovative policies to address the impact on local distribution systems from rooftop solar panels. Some concepts, such as the proposal from China's state grid operator for a global high-voltage network, will require extending engineering and operational standards across sometimes adversarial borders. It's always windy or sunny somewhere, so building a large-scale grid will help bring that renewable energy from far-flung places to load centers in major cities.

Tomorrow's T&D system also will require grid-scale energy storage – and here's where some of the most significant engineering advances still loom. Recent government research and development for batteries has been welcome, but it needs to be extended to other forms of energy storage as well. Compressed springs, spinning flywheels, ultracapacitors, chilled water, molten salts, pumped hydroelectric and compressed air energy storage all offer different performance tradeoffs. The expertise of mechanical engineers and materials scientists is sorely needed to help develop and optimize those storage concepts.

Building Better Engineers

The Paris Agreement gives unambiguous direction to mechanical engineers: Develop better hardware, algorithms and control systems to decarbonize the power sector. But the profession needs to do one other thing better. We need to build better engineers.

Because the scale of society's challenges have grown, the engineer of the 21st century will need to be more globally cognizant. And to work in an energy system that is more closely intertwined with policy frameworks, those engineers will need more policy (and political) savvy.

That means our educational approaches need to change to prepare those engineers for their careers. It seems senseless to use textbooks and chalkboards – tools we have used for hundreds of years, since before the advent of the modern engineering degree – instead of interactive, updateable digital textbooks that could enhance problem solving. We need more immersive experiences, more hands-on labs, and more exposure to real-world devices.

Just as importantly, we must provide multidisciplinary training and require graduating engineers to have the capacity to speak normal language to the array of stakeholders they will face. It won't be enough to just find the solutions; tomorrow's mechanical engineers will need to drive the conversation.

Decarbonizing the power sector isn't a job most mechanical engineers asked for, but it is the one we now have. In the process of accomplishing that task, there are a great many benefits that will certainly result from it: for example, generating electricity with greater efficiency, building a grid that is more robust and flexible, and preparing engineering graduates to have a larger impact on society. If we energize the engineering community to embrace the challenge, then when engineers look back decades from now, decarbonizing the power sector will have seemed an obvious outcome and improvement for society. And we will wonder why we didn't start it sooner.

BATH, ENGLAND

Chapter 4

Safety Needs II: Energy & the Environment

This Earth Day, Pandemic Offers Opportunity to Fix Our Air and Water Woes

AUSTIN AMERICAN-STATESMAN, APRIL 2020

> *This period of sheltering in place has reminded many of us what clean air and water look like. So how do we keep the environmental benefits without the economic destruction?*

THAT THE 50TH ANNIVERSARY OF EARTH DAY IS OCCURRING during the COVID-19 pandemic should encourage us to emerge from our isolation with a renewed focus on protecting the environment.

The worldwide quarantine has reduced our energy consumption, prevented significant water pollution and cleaned our skies. Skylines and vistas are crisp for the first time in decades as garaged cars and idled factories allow the air to clear. Cities are quieter without the piercing sounds of automobiles and traffic accidents.

This period of sheltering in place has reminded many of us what clean air and water look like. So how do we keep the environmental benefits without the economic destruction?

We do so by reinventing our urban landscape to include more greenspaces and changing our daily 8-to-5 commuter mentality.

The influenza pandemic of 1918-1919 teaches us that society can endure all sorts of challenges. It isn't easy, but it often leads to better outcomes. After the combined effects of the 1918 influenza pandemic and World War I, for instance, there was a shortage of male labor. Consequently, women joined the workforce in unprecedented numbers, ultimately helping lead to suffrage for white women shortly thereafter.

Today, one of those positive outcomes should be a new commitment to clean air and water. If teleworking retains popularity once society opens again, it could durably reduce commuter traffic, emissions and peak travel.[16] And if those who commute are more flexible with their arrivals and departures for work, we can smooth the flow of traffic, reducing congestion and making our drives more pleasant. With fewer cars and less traffic congestion, we will need fewer lanes, parking spaces and garages, opening opportunities to keep our streets clearer for bicycles and alternative transportation modes or to replace them with greenspaces so residents can maintain safe social distance in urban centers. Switching to electric vehicles will let us keep air quiet and clear even as cars return to the streets.

We also need to think big.

Nearly a century ago, after a global pandemic and stock market crash preceded a long period of economic malaise, the United States

---

16 Indeed, two years into the pandemic, it seems that remote work is likely to continue for a significant fraction of the workforce. Before the pandemic, about 4% of workers primarily worked remotely in the United States. After peaking at 50% remote work during shutdowns and quarantines in 2020, it has dropped but to a much higher plateau than prior to the pandemic. See for example "Remote Work Is Here to Stay and Will Increase Into 2023, Experts Say," by Bryan Robinson, *Forbes*, Feb. 1, 2022.

turned to large-scale renewable projects and long-haul transmission lines as a pathway forward. With the COVID-19 pandemic and recent stock market collapse, we should do the same thing again. This time, instead of building large hydroelectric power plants in the Tennessee River Valley or along the Columbia River in Washington state, we should build large wind, solar and geothermal power plants everywhere we can. And just as we launched rural electrification during the Great Depression, we should build an updated national network of transmission lines to bring renewable power from its remote locations to our city centers.

The evidence that people exposed to air pollution have a much higher risk of mortality from COVID-19 should be overwhelming incentive to clean our skies once and for all.

We need to support initiatives that improve our energy and transportation infrastructure. Large consumers with coal-heavy fleets such as the U.S., China and India need to accelerate the retirement of dirty, uneconomic coal plants and replace them with cleaner options such as wind, solar, nuclear, geothermal and natural gas. While traffic is low, cities worldwide should paint and stripe streets for dedicated bus corridors and protected bike lanes, create pedestrian-only throughways, and expand the network of electric vehicle charging stations.

This Earth Day, let's remember that although the pandemic is scary and unsettling, it also presents an opportunity for us to come together to solve difficult problems in a lasting way.

Michael E. Webber

Turning Around Our Priorities

MECHANICAL ENGINEERING, MARCH 2020

We have a few things backward that we should reverse for energy and climate.

THERE'S AN OLD JOKE ABOUT HOW WE PARK ON THE DRIVEWAY and drive on the parkway. But it isn't just the English language that gets things backward every once in a while: There are decades of energy and transportation policies that create incentives that run almost diametrically counter to what we need to accomplish this century.

What do I mean by backward policies? Let me give an example.

In the United States, we have spent trillions of dollars building a road network that is extensive, easy-to-use, and free to drive on. At the same time, we turned our mass transit system – which used to be one of the world's finest – into something that's limited, hard-to-use and expensive. While there are certainly upsides to having a coast-to-coast superhighway system, urban planning based on a large network of free roads encourages sprawl. And that sprawl brings with it time lost in traffic, wasted energy, pollution and significant expense.

We should do it the other way around.

Mass transit on narrow rights of way, such as bus lanes or railways, ought to be free, convenient, and easy-to-use. Such a change would require not only eliminating fares (already being done in places such as Kansas City, Mo., and Lawrence, Mass.) and laying down track, but also redesigning cities around people, with wider sidewalks, narrower automotive corridors, safer pedestrian crossings, and much less space devoted to public parking. At the same time, single occupancy vehicles driving on expensive, expansive highways should be costly enough through automated tolling to cover the full cost of their impacts, which includes air pollution, climate emissions, noise pollution, congestion and maintenance. The current model uses taxes rather than user fees to pay for roads, which means we are all subsidizing those emissions and sprawl.

One way to think about it is that packed roads and empty trains and buses are failures of our transportation system, so we should implement policies that clear the roads and fill the trains. But of course, first those trains need to exist.

Federal energy policy is backward too. The official position is that we should strive for "energy dominance." But instead of encouraging production and discouraging consumption to increase the amount of exportable energy, tariffs on energy technologies such as solar panels and rollbacks on efficiency standards take us in the opposite direction.

When it comes to energy policy, however, the biggest failures have to do with pollution and climate change. Our current policies have us locked into outcomes that run counter to what most people want.

Companies that produce solid or liquid industrial waste, whether it's toxic chemicals or noxious wastewater, expect to pay disposal costs. But as it stands, companies can emit greenhouse gases for free – they don't have to pay a price to use our common atmosphere as their dump. Conversely, companies that want to avoid the lasting damage of their CO_2 emissions often have to invest in expensive technology to capture their emissions or switch their equipment to cleaner versions. Such spending puts clean companies at a competitive disadvantage against the ones that pollute freely.

This mismatch between what we need (making it expensive to pollute) and what we have (it's free to pollute) is the primary reason why we're in this climate mess in the first place.

So how can we turn around these backward policies?

One way is to put a price on carbon emissions. Even better, we could pay carbon cleanup companies to remove CO_2 from the atmosphere, just like we pay municipal garbage companies to pick up our trash from the curb and we pay wastewater treatment companies to clean up our sewage. Instead of paying energy companies for carbon they extract from ancient reserves below ground, we should pay a premium for carbon they harvest from the atmosphere.

From the perspective of today, where we allow carbon to flow from the ground and into the air, a system where carbon flows the other way – or one where it circulates with zero net emissions – will require new thinking. Reversing long-standing policies can be a tricky task, but the first step in doing so is to admit we have things backward in the first place.

The Oil Industry Is Part of the Solution

MECHANICAL ENGINEERING, FEBRUARY 2020

Activists want to punish the petroleum industry, but that industry has the technological chops to tackle climate change.

HUMANS ARE HARD-WIRED FOR FAIRNESS. WE FEEL SATISFIED when everyone gets their just reward and are outraged when we discover cheating. But that innate desire for justice can get in the way of solving our biggest challenges.

Take climate change: When scientists and environmental activists take stock of the mess we are in, the oil and gas sector is a handy villain. People tapping into their instinct for retribution say the petroleum industry ought to be punished for the damage it has caused and cut out from any opportunity to participate in the upcoming transition to a clean-energy economy. To activists who have made climate change a top priority, anything less feels like inviting an arsonist to help put out the fire.

As with everything, however, the truth is more nuanced.

If tackling climate change is something we want to do quickly and with as little social disruption as possible, then the oil and gas industry is, in fact, a critical partner. Petroleum companies have some of the deepest pockets and most technically capable workforces around.

Is there a way to work with them, rather than against them, to promote a low-carbon future?

Unquestionably, many oil and gas companies have been bad actors. At best, the petroleum industry has ignored the problem while making a profit off the products that worsened the situation. At worst, it actively worked to delay action by funding misinformation campaigns or lobbying to delay policy action.

But blaming the industry leaves out our own culpability for our consumptive, impactful lifestyles. Oil consumption is as much about demand as supply.

Rather than finding someone to blame, let's look for who can help.

My friend Mark Brownstein, senior vice president at the Environmental Defense Fund, once told me, "Sometimes people are afraid of solutions at the scale of the problem." We latch onto easy, rinky-dink, boutique, niche solutions – reusable drinking straws come to mind – that don't move the needle but make us feel better about ourselves. They are the policy equivalent of pet rocks.

To make a real difference in battling climate change, we need solutions that work at scale. For instance, enhanced geothermal energy can draw industrial-scale heat from deep underground. Biogas can convert detritus such as manure and food waste into renewable methane at large scales. Perhaps the most critical carbon solution – and one that must be enacted at the largest scale – is CO_2 capture, utilization and sequestration.

Those solutions require companies with a particular set of expertise – expertise found in abundance in the oil and gas sector.

Oil and gas companies have the drilling expertise we need to reach geothermal resources. And the infrastructure and capabilities for responsibly capturing large volumes of CO_2 and handling it for sequestration. The industry knows how to separate CO_2 from methane, which is useful in biogas purification.

In addition, petroleum companies have hundreds of thousands of engineers and scientists with great capabilities and supply chains that span the globe. They have refineries and other petrochemical facilities that – like magic – can convert materials from one form to another. That capability will be needed to repurpose CO_2 captured from smokestacks or the atmosphere into economically valuable products, building materials, fuels or chemicals.

But we need to get those companies on board. Today, there is little to no penalty for emitting greenhouse gases and precious little economic incentive for restraining them. Flipping the incentives to unleash the innovative spirit of this pioneering sector will require policy changes at the local, national and global level. Even some oil and gas companies like Shell have admitted that they will only step up into leadership roles if they are given clear signals, such as bans on diesel engines or petroleum-powered cars altogether like some cities (Paris), states (California, New York and Washington) and countries (Norway) have recently implemented.

Ultimately, we need every tool in the toolbox. To solve the climate crisis quickly and with the least amount of pain will require all hands on deck – especially those experienced, capable hands from the global oil and gas sector.

How Cheap Gasoline Hurts the Environment

THE DALLAS MORNING NEWS, JUNE 2016

> *A sustained era of low gasoline prices means it will be harder to encourage conservation or to invite alternative transportation fuels such as biofuels, electricity or natural gas into the transportation sector.*

THE SUMMER DRIVING SEASON IS UPON US. AND BECAUSE OF A sustained period of low oil prices, drivers will most likely enjoy substantial financial benefits of cheaper gasoline prices all summer long.

But although low oil prices are a boon for drivers, the situation is a little more complicated for policymakers.

Even though oil prices today are hovering at the same level as they were a year ago, competition among refiners and the time lag between oil prices and gasoline prices have passed. That means drivers in 2016 enjoy gasoline about 40 cents cheaper per gallon than drivers the prior year did.

Industries that depend on energy as an input – such as chemical manufacturers, factories or smelters – will enjoy higher profit margins or have more money available to invest in new equipment. Homeowners have similar benefits. Lower energy bills are like a pay raise and a tax cut, giving us more money in our pockets that we can use to pay for elaborate summer trips.

Americans have received the message. After driving a record 3.1 trillion miles in 2015, drivers are projected to hit the roads even more while taking more summer trips and spending more on vacations. The U.S. Energy Information Administration recently announced that gasoline consumption in March 2016 set a record, an early sign that our path to driving more miles is already underway.

But cheap gasoline also has some real challenges, especially for the energy sector.

Hundreds of thousands of layoffs at oil, gas and coal companies mean that people in some regions will be driving not for summer vacation, but to look for jobs elsewhere. Billions of dollars of capital investments that kicked off the shale boom have slowed down, creating a drag on the stock markets. It turns out that oil price increases, which make life expensive for the typical working and commuting U.S. worker, can be just as difficult for the economy to accommodate as oil price decreases, which rattle the markets and bankrupt companies.

And, with current policy priorities such as decarbonizing the economy and reducing oil imports, low gasoline prices are confounding. Before last year's new record, national vehicle miles traveled peaked in 2007 at just over 3 trillion. Between then and 2014, higher gasoline prices encouraged people to consider more fuel-efficient vehicles or alternatives such as mass transit. That means we were driving less, and when we drove, our cars required less fuel per mile. Both of those were contributing factors to decreasing carbon dioxide emissions, and, coupled with increases in domestic production, also caused oil imports to fall.

But with cheaper gasoline, we're again buying bigger cars and driving more, and our domestic production is leveling or dropping. That means increases in carbon dioxide emissions, gasoline consumption and imports. All of which feels like a step backward compared with the steady progress we made for years with our energy, security and carbon dioxide policies.

A sustained era of low gasoline prices means it will be harder to encourage conservation or to invite alternative transportation fuels such as biofuels, electricity or natural gas into the transportation sector.[17]

Although the economic benefit of cheap gasoline is nominally a good thing, in the modern policymaking era, where economic interests need to be balanced with security and environmental priorities, it is a conundrum for policymakers.

But there are some things that can be done. First, we should keep pushing for more research and development, particularly for more fuel-efficient cars and alternatives such as electric vehicles that make transportation more efficient. Second, we should remain vigilant with our fuel economy standards so that on average we continue to buy fuel-efficient cars no matter the price at the pump. Third, we can keep pushing for a culture that is mindful about conservation and efficiency.

The summer driving season is a fun time to escape with the family. It's also an opportunity to put some of these more efficient options to work.

---

17 It is notable that oil and gasoline prices were very high in 2021 and 2022 because of a strong economic recovery, the Russian invasion of Ukraine, and reduced investments by the oil and gas industry in exploration and production during the early months of the COVID pandemic. Those higher prices have helped facilitate accelerating adoption of electric vehicles. See for example, "Electric Vehicles Start to Enter the Car-Buying Mainstream," by Jack Ewing and Peter Eavis, *New York Times*, Nov. 13, 2022.

Include Agriculture in Emissions Policy

CORPUS CHRISTI CALLER TIMES, MARCH 2015

A comprehensive climate policy needs to do more than tackle smokestacks. It also needs to do something about agriculture.

IN LIGHT OF PRESIDENT BARACK OBAMA'S LOOMING CARBON regulations for existing U.S. power plants, it's worth remembering that a comprehensive climate policy needs to do more than tackle smokestacks. It also needs to do something about agriculture. And more broadly, Texans and the rest of the nation need to think more environmentally about the way they eat.

After fossil fuel combustion, agriculture is the second-largest emitter of greenhouse gases in the nation. Despite consuming 2 percent or less of our energy, agriculture generates 10 percent of our emissions. And while other sectors of the economy are reducing emissions, agriculture is heading in the opposite direction. This trend is bad for statewide and national efforts to reach net-zero emissions.

Agriculture primarily emits two potent greenhouse gases, nitrous oxide and methane, from activities such as the application of nitrogen-based fertilizers, as well as manure management and burps from cows. In addition, agriculture is a source of dust and precursors for air pollution. But agriculture often gets a pass when it comes to air-quality laws. Historians can debate why this happened in the past, but to allow it to continue by giving agriculture a pass on its greenhouse gas emissions would be a huge mistake moving forward.

Regulations on carbon emissions disproportionately affect states that rely heavily on coal-fired electricity such as Indiana and Illinois, while states such as Washington that rely on hydropower would not be noticeably affected. By contrast, everyone in America eats, so putting a price on the carbon intensity of food would spread more uniformly across society so we all share in the benefits and costs.

It's true that farmers will have to make adaptations, but through incentives, we can make these revenue-neutral for farmers who lower emissions. If we were to put a price on agricultural carbon, consumers would face higher prices for more carbon-intensive foods, such as meat. The change would encourage shifts in diet and could dramatically lower emissions because meat is known to be much more carbon intensive to produce than fruits, grains and vegetables.

The agricultural sector may balk, saying that holding it accountable for emissions the way we hold other sectors of the economy accountable will be bad for business. But this is not true. There are plenty of ways farmers can adapt to a lower-carbon world and even find new revenue streams in the process.

Consider the 100 million tons of manure that livestock generate each year. Those piles are a major source of greenhouse gases and a major headache for farmers, who have to deal with economic, environmental and legal liability from odor and handling costs. Those same mounds of manure, however, are potentially a rich source of biogas, which could offset 4 percent of our annual natural gas consumption. This might be one of the easiest, cheapest and fastest ways to produce a significant amount of renewable, low-carbon, domestic energy that is available around the clock.

Another big opportunity is to reduce food waste. Amazingly, 25 percent or more of our food is wasted, which amounts to a tremendous, equally wasteful use of energy and emissions. We quite literally throw that food-energy right into the garbage. Reducing food waste is a straightforward way to reduce energy and emissions from the food system, and it should save money for everyone along the entire food supply chain, from farmers to retailers to grocery shoppers.

Most important, certain land management techniques can sequester hundreds of millions of tons of carbon into soils each year. No one is better suited to do this at larger scale than the agricultural sector. Putting carbon back into the soil through conservation programs does society an important service, and farmers should be paid handsomely for it.

The time has arrived to tackle climate change in a comprehensive way. At a policy level, we have to stop giving agriculture a free pass. If we can drive more-efficient cars, insulate our homes and use less coal, surely we can also reduce emissions from the food we eat.

Section III

Connection Needs

SEA ISLAND,
GEORGIA

Chapter 5

Connection Needs I: Energy & the Global Economy

An Invisible Hand Driving Energy Policy

MECHANICAL ENGINEERING, APRIL 2020

> *The finance and insurance industries have a major say in how we address mobility and climate issues.*

ENGINEERS LIKE ME TYPICALLY BELIEVE WE HAVE THE ANSWER TO everything, and that better technology is the path to a better future. But look at the major revolutions in energy and mobility, and you'll see another force playing a bigger role than most of us realize: finance.

While the Model T was a clear breakthrough of production, Henry Ford wanted customers to save up the full purchase price and pay with cash. The car industry as we know it was made possible thanks to GMAC, GM's financing arm that just over a century ago opened up individual auto ownership via car loans. And automobile insurance made sure that owners of these expensive and dangerous machines wouldn't be bankrupted in case of an accident.

Finance has shaped the development of the capital-intensive oil industry too. Wildcatters hunted for money just as aggressively as they hunted for oil. Without the money, they could not afford to pay for their workers or expensive equipment to drill prospective wells. Relatively easy access to cheap capital after the global financial crisis of 2008 helped unleash the shale boom of growing oil and gas exploration and production in the United States.[18]

In the power industry, the cost of money is a major factor in what kinds of power plants or other infrastructure can be built. Meanwhile, clean-energy programs can speed the adoption of rooftop solar panels by shifting costs to homeowners' property tax bills rather than expensive bank loans.

Finance and real estate will continue to be forces to contend with. But in the coming decades, I believe the insurance industry will shape mobility and energy policy like never before.

For instance, the insurance industry likely will drive the adoption of electric and self-driving vehicles. Peer-reviewed research in a 2018 volume of the journal *Transport Policy* found that insurance premiums for electric vehicles could be as much as 35 percent cheaper than for equivalent gasoline models. What's more, because robot drivers will be safer than distractible humans, the insurance for an autonomously driven car could be cut by nearly half. This differential will push ordinary car owners to buy clean and safe autonomous electric vehicles, leaving what we now think of as "conventional" automobiles as the playthings of the rich.

Actuaries could drive climate action, though indirectly.

---

18 Notably, more expensive and difficult-to-access capital for oil and gas companies in 2022 because of higher interest rates and decarbonization mandates for investors might do the opposite, namely lead to less exploration and production for oil and gas.

Parts of California are already uninsurable because of fire risks. Losses due to wildfire damage have reached billions of dollars per year, forcing insurers to raise rates to cost-prohibitive levels – or to cancel coverage altogether. This, in turn, impacts property values.

It's not just wildfires: Floods also loom large. Miami is both sprawling and low lying, meaning that even throwing up a short wall against storms and sea level rise will cost billions. New York City is also considering constructing a barrier to protect some or all its extensive shoreline, but that 25-year project would cost more than $100 billion to implement.

The alternative – waiting for the storm to hit – cost Houston nearly as much in cleanup after 2017's Hurricane Harvey. Much of the damage wasn't insured, and in the face of rising seas and more severe storms, it is likely insurance companies will increase premiums or stop providing coverage in low coastal areas.

Just as with electric cars, insurance costs and availability will start to force technology and infrastructure choices. For example, the high up-front costs of burying power lines – important to reduce fire risk – might be justified by lower premiums for utilities and for at-risk property in their service areas, since they will no longer be threatened by fires sparked from overhead lines. More broadly, insurance costs will push people to abandon floodplains or coastal cities and encourage infrastructure builders to prepare for a hotter world by protecting properties and making their systems safer.

In other words, while policymakers and individuals roundly ignore technology and climate experts and plan on business as usual, the apolitical and unemotional insurance industry – ignoring slogans and considering only the numbers – will force them to eventually take climate change seriously. Compared to the costs of sea walls, infrastructure hardening and a potential retreat from the coasts, the price tag for decarbonizing the energy sector seems like a bargain.

Michael E. Webber

Over the Hills and Far Away

MECHANICAL ENGINEERING, DECEMBER 2021/JANUARY 2022

> *With global supply chains locking up, it's worth looking to the energy industry for some lessons.*

THE WORLD'S ECONOMIC LEADERS THOUGHT THEY HAD ALL BUT perfected the just-in-time supply chain on the eve of the pandemic in 2020. No one would say that now. Over the intervening months, we've seen how fragile it was all along: One container ship stuck in the Suez Canal sent global markets into turmoil, and a backlog of cargo deliveries at the Port of Los Angeles now makes these same leaders fret.

Supply is a challenge that the energy industry has long faced. Energy resources such as oil and natural gas can be extracted in one far away country and show up at our doorstep in another like magic. The fact that the large oil companies that dominated the post-World War II energy landscape were known as the "Seven Sisters" only underlined the fairytale aspect, since many European folktales open with lines such as "beyond seven lands and seven seas."

Yet, for all the ingenuity (and political finesse) involved in the global petroleum trade, the energy industry also realized that it left consumers in a vulnerable position if they lived in countries with different foreign policy aims than those of producer countries. That is why local production is so crucial.

Up to now, however, local energy production has been prone to rapid depletion. In the old days, this would have been firewood from the trees out back. In 19th-century America, growing population far outstripped the land's ability to reforest, and the Eastern United States today has very few stands of virgin, old growth forests.

Coal replaced wood, at first from subsurface mines close to population centers. Once those depleted, we turned to large open-pit mines far from population centers in Wyoming and elsewhere. The U.S. oil industry began with seeps in western Pennsylvania, and gas was initially manufactured from coal in city gasworks. Today, pipelines for liquid and gaseous fuels like crude oil and natural gas allow our energy supply chains to sprawl across the continent.

The national energy supply chain isn't perfect – its infrastructure can be vulnerable to natural and manmade disaster, cutting fuel and power when it is needed – but it is surprisingly robust. Part of that is due to the capacity built into the system. The electrical grid is sized to serve the projected need on the hottest day of the year rather than providing "just enough."

A lot of it is also an outcome of the energy system's sense of space. Local energy, available when you need it, is good; less dependable local energy that can be stored is next best. Perhaps best of all is local energy that unlike nearby woodlots or coal mines, never depletes.

The wind that flows past our homes and through nearby turbines, the photons that hit our rooftops, and the waste we convert on-site into heat or power: These technologies at local scale are more expensive than remote power plants, but they never run out or face embargo and they also avoid the cost – and risks – of moving energy long distances.

This "think global, act local" mindset has begun to be transformative for the energy sector by facilitating the growth of distributed renewables while reducing risk to supply-chain disruptions. This is a good thing, and something the global manufacturing supply chain should study.

Let me end on a note of caution. Switching away from those hydrocarbons to a renewably electrified future is appealing because it reduces the security risks of today's fuel mix. But in the big picture view, replacing strategic liquids and gases with critical minerals such as lithium or tungsten just shuffles the countries we worry about. Instead of fretting about oil from the Middle East or gas from Russia, we may soon worry about lithium from Bolivia and arsenic from China, two countries whose relationship status with the United States is best described as, "it's complicated."

As in many fairy tales, getting your wish granted doesn't always lead to happily ever after.

Boomtown

ALCALDE, JULY/AUGUST 2014

After fears about peak oil, greenhouse gases and our dependence on Middle Eastern oil, the world of energy has suddenly flipped. Oil and gas production are up, imports and CO_2 emissions are down. The shale boom is giving America new confidence – but is it all it's made out to be?

FRACKING IS DOWNRIGHT CONFUSING. WE CAN'T EVEN AGREE HOW to spell it. The *Wall Street Journal* and other leading media outlets spell it with a "k," while industry today prefers the clumsy, tongue-twisting "hydraulic fracturing." The irony of this preference is that the term "fracking" was initially used by industry as convenient shorthand until it was co-opted by the environmental movement as a handy pejorative due to its similar structure to a common vulgarity. Signs of "Don't Frack Our Land" and similar complaints are a common maxim among opponents to fracking. So industry has gone to the less-efficient and more-scientific "hydraulic fracturing" as a way to distance itself from the prevalent f-bomb. The meaning of "fracking" varies depending on who's using it. For the purpose of this discussion, when I use "fracking," I mean hydraulic fracturing in shales with horizontal drilling.

This disagreement over something as simple as how to pronounce and spell the technique is a sign of just how entrenched and polarized the outlooks are for the shale revolution. And, like other earlier energy revolutions, the story has a particularly Texan flair to it.

Is shale production an economic miracle, or is it leaving some people and industries behind? Is fracking an environmental disaster or will it save the planet? It is hailed as the liberator from our dependence on foreign oil, but does it threaten to bind us even closer to the geopolitical volatility of international energy markets? The truth is a little more nuanced than shale promoters and detractors imply.

The Texas Connection

The shale revolution was Texas-led and Texas-bred. In fact, it's so new that its history is still being written. Russell Gold, the senior energy reporter for the *Wall Street Journal*, just released a definitive account of the history of fracking with his book *The Boom: How Fracking Ignited the American Energy Revolution and Changed the World*. One of the first lessons Gold reveals is that the new techniques of fracking are not new at all. Fracking in one form or another has been deployed since the 1860s with a variety of approaches, including dynamite and even nuclear detonations. The great enabler for the modern version of shale production was the combination of hydraulic fracturing, horizontal drilling and sophisticated fluids that include a variety of chemicals. But even those techniques had been used independently for decades. The idea that shale would contain oil and gas is not new either, as geologists have known for a long time that these formations – the source for major oil and gas reservoirs – hold a lot of the desired materials. It was just technologically and financially prohibitive to extract them.

But then a Texas oil and gas producer, George Mitchell, cracked the code for making shale a worthwhile endeavor. With stubbornness, persistence and a capable team, he achieved success after decades of attempts that would have made other experimenters give up, and so now we're off to the races. In many ways Texas is the epicenter of the worldwide energy revolution. The boom is on. Thankfully – or possibly regretfully – this is familiar territory for us.

When the original Texas energy revolution symbolically and literally erupted with Spindletop more than a century ago, it put the world on notice that instead of depending on territories under the heavy hand of Russia, Texas could be the energy salvation. At the time, Azerbaijan was by far the world's largest oil producer. The Nobel brothers, of Nobel Prize fame, earned a significant fortune there enroute to endowing their famous award for scholarly breakthroughs. As a counterweight to

Russian control over oil and gas, Texans happily exported their oil and expertise for a nice profit.

When this latest Texas revolution broke onto the scene a few years ago in the Barnett Shale outside Fort Worth, it put the world on notice once again. Since more than half of Russian gas exports to Europe pass through Ukraine, it might serve Europe – and other countries across the globe – to consider Texas as an alternative source of gas, or as a model for how to produce your own oil and gas.

The good news is we have major shale formations in production throughout the United States. There is an old adage in the oil industry that Jack Randall, a successful oil and gas veteran, likes to share. If someone asks, "Where's the best place to find oil?" the answer is, "The same place where it has been found before, just drill deeper and try a little harder." Such is surely the case in the Permian Basin in West Texas, which has come alive again with the hustle and bustle of another oil boom.

In Texas we love our oil booms. Pennsylvania, which hasn't had an energy boom since its oilfields and anthracite mines played out in the late 1800s, is experiencing its own version of an energy boom today with the Marcellus and Utica Shales. Booms aren't new and we know this boom will eventually end. There's a famous bumper sticker in Texas that reads something like this: "Dear Lord, please give me just one more boom and I promise not to piss it all away next time."

Fracking Saves the Economy ... Except When It Doesn't

Fracking provides a much-needed boost to the economy, especially in producing states. Energy companies that are busy drilling will hire more people and enrich landowners who hold mineral rights. That economic benefit ripples through the economy to busy restaurants, higher wages and lower unemployment.

It is also helping to keep our energy prices at a reasonable level. The flood of domestic oil and gas production combined with bottlenecks – both real and political – inhibit the movement of energy out to global markets and as a result, domestic prices are usually lower than international prices.

The shale story has affected natural gas too. The abundance of natural gas and scarcity of export terminals means we have a domestic glut that pushed prices far below market prices in Europe and Japan, so by comparison our energy is incredibly cheap. U.S.-based chemical companies, many of them lining the Gulf Coast in Texas, often use natural gas and natural gas liquids as a feedstock for making plastics, fertilizers and other materials. That gives them a remarkable price advantage over chemical companies around the world that pay international natural gas prices or use petroleum as their feedstock. Consequently, there are tens of billions of dollars of announced investments in chemical manufacturing capacity in the U.S.

Trevor Houser's and Shashank Mohan's book *Fueling Up: The Economic Implications of America's Oil and Gas Boom* notes that the oil and gas boom, which accelerated around 2009 and 2010, came at a very fortuitous time. Lower energy prices are like a tax cut, putting more cash into people's pockets, and the new investment in oil and gas production has been like a stimulus package. Those combined effects have a cumulative impact on the economy that should last about a decade.

Ok, so what's the catch? As the oil and gas sector takes off, the competition for labor and capital drives up the costs for both. Domestic resource booms tend to make the dollar stronger, which hurts exports. That means other export-oriented industries that don't care much about energy – like automotive or electronics manufacturing, where labor and capital costs are much more important and where a weak dollar is preferable – will be paying their employees more, will have more expensive debt, and will be exporting against the headwind of a dollar that's been made stronger by oil and gas.

That inflationary pressure on prices doesn't only hurt industry, it also hits the people in boomtowns who aren't landowners, business owners, or somehow working in oil and gas. Teachers, firefighters, and other workers all of a sudden find themselves having to pay more for milk (at least for a little while), or getting priced out of their apartments to make room for oil and gas companies who will pay top dollar for employee housing. In many cases, they are collateral damage from the boom.

Success for oil and gas comes at the expense of other forms of energy: Coal, nuclear and renewables are all having a tough time competing against cheap natural gas, for example. Texas is the leading state for natural gas, but it is also the leading state for wind power, the largest consumer of coal in the nation, and the site of the nation's largest nuclear power reactors. While we're an oil and gas state by identity and history, when push comes to shove, we're really an energy state. That means we care about the fate of the other fuels too.

Rural Revitalization

One of the nice things about this oil and gas boom is its rural flavor. In many ways the shale revolution has been a useful on-ramp for rural revitalization. *Bloomberg Businessweek* reported how the Bakken shale boom flooded rural banks with cash. Local tax receipts are up, which gives small communities badly needed money to build and improve schools and courthouses.

But there are some unexpected outcomes. Formerly poor school districts in Texas that received a lot of money from the state are richer. While the local government budgets are higher, so are the expenses. Counties grapple with rural towns filled with broken roads, and regional hospitals are fuller than normal, both casualties of the flourishing oil and gas activity.

These costs are in contrast with the benefits of bustling restaurants, retail stores with increased sales, full hotels, and every other economic activity you can imagine. While the stereotypes of an all-male industry feel dated– oilfield work trailers have more women in them than ever before – it is still the case that rural strip clubs have never done better, and that's the rub: These small-town communities, many of which are very religious and socially conservative, are often a destination for people who want to avoid urban problems. But then their little hamlet starts to attract thousands of men, drugs, prostitutes and bar room brawls, along with flowing dollars. Managing these tradeoffs moving forward is one of the grand challenges for boomtowns.

The Environmental Conundrum

Fracking is the environmental solution that environmentalists love to hate. Fracking has proved perplexing for the environmental movement, dividing its factions into disparate camps: Those who oppose fracking because it still extracts a fossil fuel that emits CO_2, and those who support fracking as the fuel that will enable a graceful path from coal and oil toward a future based on renewables. It's a source of noise pollution, light pollution, air pollution, land pollution, water pollution and greenhouse gases. The equipment makes a racket for neighbors. The lights from flares and equipment can be seen from space. Gases escape from production sites, raising concerns about their contributions to climate change. The drilling pads, usually a few acres apiece, mar the landscape. Flying into the Dallas-Fort Worth airport gives a bird's-eye view of a patchwork of thousands of production sites that have been cleared of vegetation and tamped down. They look like misplaced, rectangular, white puzzle pieces scattered over the land as far as the eye can see.

Fracking also propagates the fossil fuel era longer than many environmentalists want, which means more decades of CO_2 emissions and worsening climate risks. Add in the risks associated with fugitive emissions of methane, a very active greenhouse gas, and things look even bleaker to the environmental community.

Yet natural gas is the prince of fossil fuels: It is cleaner than petroleum and coal. It can cut our pollution of SOx, NOx, particulate matter, and heavy metals like mercury dramatically, in some cases by more than 99 percent. And despite the methane leaks and other concerns, cheap natural gas liberated by fracking has enabled a rapid decarbonization, as we displace coal with natural gas in the power sector. In fact, while carbon dioxide emissions continue to increase globally, U.S. emissions have dropped below 1990 levels.

An issue that makes fracking even thornier is water use. This quote about a boomtown captures some of it: "Many's the time you'd see a man come in, order a quart of whiskey poured in a bowl and go to washing his face and hands. Damned good reason for that: Water had to be hauled miles and it cost like blue blazing hell ... it took many hundreds of gallons of water to drill a well ... but water cost three dollars a barrel." This anecdote reveals the value of water to oil and gas operations and how boomtowns have an inconvenient knack for being located in regions enduring severe drought, forcing operators to bring in water from far-flung locations. The funny thing is, that quote is from a 1939 story in *Cosmopolitan* about Burkbunett, a boomtown in Texas. That article inspired the movie *Boom Town* starring Clark Gable and Spencer Tracy. The water tensions for the oil and gas industry in Texas are hardly new.

Water is a flashpoint for many people who do not want to compete with energy companies to get the water and are worried about contamination of their aquifers or surface spills that will get into the waterways. There is good reason for this concern, as a typical fracturing job requires anywhere from 2 to 9 million gallons of water per well. That means a lot of truck traffic moving water to the well pad and the risk of shortages or higher prices for other users. Those same wells also return millions of gallons of wastewater laced with the chemicals that are injected to enhance productivity, along with the naturally occurring chemicals in the shale that come to the surface. The salt levels of so-called produced water make the ocean look like freshwater by comparison.

While these concerns and risks are very real – after all, the word "hydraulic" in "hydraulic fracturing" means "water" – there are some surprising ironies. It turns out that we use water to produce every form of energy. While the steady stream of water trucks is an obvious indicator of the water needs for fracking, biofuels from irrigated corn for ethanol are about 100 times more water intensive. And, despite the additional water used with hydraulic fracturing to produce natural gas from shale formations, University of Texas Austin research in engineering and geosciences has revealed that natural gas use saves water over its entire life cycle because natural gas combined-cycle power plants have less than half the water intensity of coal plants. Shale gas is leaner from a water perspective than people anticipate and is a drop in the bucket compared to irrigated agriculture. That might not be much comfort in some shale plays, as the localized water impacts from shale gas extraction can still be significant, and the water savings at the power plant might occur elsewhere several months later.

For many people, the issue of water quality is worse than quantity. While some of the accusations by fracking opponents are overblown, it is important for oil and gas producers not to sidestep the fact that there are real risks to water quality from oil and gas operations. It is also important for stakeholders to realize that those risks are not specific to fracking, rather they are present wherever oil and gas, or really any energy operations, are prevalent. Recent high-profile events such as deepwater blowouts, coal chemical spills, coal ash spills, and mountain-top removal mining that buries hundreds of miles of waterways serve as reminders that fracking is hardly the only form of energy production that puts water at risk. Even the renewables are not squeaky-clean: UT-Austin research a few years ago evaluated how biofuels production can lead to nitrogen-laden runoff that gets into the waterways. Fracking's scorecard on water quality is still being determined, but by comparison, it isn't really that bad.

National Security Implications

The national security effects of the shale boom are just as nuanced as the economic and environmental tradeoffs. Simply speaking, the increased domestic production of oil and gas and the associated reduction in energy imports are good for reducing our national security vulnerabilities and improving our geopolitical posture. There are many reasons to cheer on this rapid turnaround, and peace activists and national security hawks can revive their ambitions to end our foreign entanglements brought about by our desire to protect the flow of oil from foreign fields to our fuel tanks.

But now the conversation has shifted.

The conversation of imports is actually ramping up – though this time with oil by pipe from Canada rather than by ship from the Middle East. It sounds better than what we worried about before, but it still complicates the conversation, and now we're talking about building more liquefied natural gas (LNG) export terminals and allowing crude oil exports for the first time since the 1970s. We can boldly tell Europe that our LNG will solve their energy security problems too, as it will let them kick their expensive habit of mainlining Russian gas via stainless steel syringes that puncture right into the heart of Western European load centers like Berlin.[19] So oil and gas enthusiasts are quick to point out that not only have we solved our energy problems, but we're about to solve Europe's security problems too. While this thought is appealing, if we start exporting large volumes of crude oil and natural gas to the world, then we are actually coupling ourselves more closely to the geopolitical volatility of global energy markets, rather than becoming more energy independent. We will still have national security implications from our energy decisions, we will just be on the other side of the equation.

The Lessons Learned

On balance, it's more good news than bad news. Whether we like it or not, fracking seems like it's here to stay. So it's worth asking how this good fortune came about. Was it just dumb luck, or was there a method to the madness that produced such a startling turnaround? In particular, is this revolution the consequence of healthy market forces, good policies, disruptive technologies or something else?

Market triumphalists like to note that capital-intensive energy markets are highly evolved and operate efficiently. When oil and gas prices were very high in the late 2000s, the markets responded by unleashing investment and expertise to bring value from what had been considered an uneconomic resource just a few years prior. This

---

19 After the Russian invasion of Ukraine in early 2022, this point became more acute. Our work at the University of Texas Austin analyzed the potential for liquefied natural gas exports to help: "The US role in securing the European Union's near-term natural gas supply," by Arvind P. Ravikumar, Morgan Bazilian and Michael E. Webber, *Nature Energy*, May 26, 2022. Source: https://doi.org/10.1038/s41467-022-34100-3

production is taking place almost entirely on private lands with private companies using private investment that was triggered by high prices.

Good government supporters point out that a steady stream of R&D investments from the Department of Energy in partnership with George Mitchell for field-scale experiments in the Barnett Shale from the 1970s to the 1990s kick-started the whole trend. The Energy Policy Act of 2005 clarified the regulatory framework for hydraulic fracturing by excluding the process from the rules that govern underground wastewater injection. That regulatory clarity and patience of government research program managers were critical ingredients that allowed fracking to take off.

Technocrats cheer the fact that disruptive technologies are changing the face of global energy production. Industry's investment in technology evolved over decades. Field-scale shale production started in the 1920s in Kentucky, horizontal drilling was first demonstrated in the 1930s, and hydraulic fracturing was first used as a well-completion process in the 1950s. Integrating all three with advanced chemical additives was a 21st-century idea whose time had come.

In fact, it is the convergence of all three – highly functioning markets; effective government actions; and disruptive technologies – that enabled the entire revolution. For the first time since the 1960s our energy markets, policies and technologies are all pointing in the same direction: up.

Michael E. Webber

A Dirty Secret: China's Greatest Imports

EARTH MAGAZINE, JANUARY 2011

> *The upshot is that as the world turns toward China to be its dirty manufacturer, we all clean up our books, pushing our emissions and energy consumption onto them. We let China produce and ship our goods, and then say, "Who me? I don't produce emissions. I've cut mine. China is to blame."*

THE U.S. AND MUCH OF THE WESTERN WORLD HAVE A DIRTY secret. While we claim to be working diligently to decrease our emissions and switch to cleaner, nonfossil fuel energies, we are just exporting emissions to other countries, most notably China. We don't talk about it. We get on our soapboxes at international meetings and claim to be making great progress to halt ever-increasing carbon dioxide concentrations in the atmosphere. And we complain vociferously about developing countries – again, most notably China – not doing the same.

Meanwhile, the industrial world is producing a smaller fraction of durable goods, and China is picking up the slack: For example, the value of U.S. manufacturing has nearly doubled in the last 20 years, whereas Chinese manufacturing has gone up by an order of magnitude over the same time span. The upshot is that as the world turns toward China to be its dirty manufacturer, we all clean up our books, pushing our emissions and energy consumption onto them. We let China produce and ship our goods, and then say, "Who me? *I* don't produce emissions. I've cut mine. China is to blame." We are misleading ourselves into believing that we're cleaning up our act.

Decarbonization at Home, Carbonization Abroad

One of the fascinating stories of the 21st century's first decade has been China's rise as the world's largest energy consumer. At first blush, this leadership is a good sign for the world, as it signals China's economic growth, hundreds of millions of people being lifted out of poverty, increased capabilities as the world's factory and prospects for democracy. But there's a downside: Compared with the United States, the world's second-largest energy consumer, and the European Union, China's energy mix includes a high fraction of coal – 70 percent in China compared with 23 percent in the U.S. and 18 percent in the EU. That makes China's energy consumption much more carbon intensive on average, and so the country is now also the world's leading carbon emitter. And its demand for energy is increasing to keep up with its economic growth.

Meanwhile, because of all the attention to carbon emissions and energy consumption, the U.S. is starting to change its behavior. Our energy consumption has dropped from 359 million Btu (British Thermal Units) per person per year in 1978 to 308 million Btu per person in 2009[20], continuing a decades-long push toward greater energy efficiency. Furthermore, we are using less coal. Just a few years ago, coal made up more than 50 percent of our electricity fuel mix; at the time of this essay writing, it was just 45 percent and dropping.[21] And with looming standards for air quality that are likely to be much stricter in the future, many foresee a steady replacement of coal plants by wind, solar, nuclear or natural gas combined-cycle power plants. For environmental advocates, this trend has been good news, because it reveals the potential for steady, albeit slow, decarbonization of the U.S. power sector. The bad news is that much of the success in reducing our energy consumption has been achieved by shifting some manufacturing to China. The worse news is that now we're not only sending manufacturing to China, we're also sending coal itself.

While the world is turning away from coal, Chinese demand keeps on rising. And because China's coal reserves and production are not sufficient, the country has become the world's largest coal importer.

---

20 Energy consumption in the United States was 293 million Btu per person in 2021.

21 In 2021, coal provided about 22 percent of our electricity, according to the U.S. Energy Information Administration.

That means American providers of coal, whose domestic markets are shrinking, are looking for ways to send their product to dirtier, less-efficient coal plants in China. And for U.S. coal producers with easy access to West Coast ports, business is booming. Coal mines in the U.S. might be just as productive as ever, even in the face of declining use at home, because of growing use abroad.

To illustrate, U.S. coal exports grew more than 50 percent in the first half of 2010 compared with the first half of 2009, from 26 million short tons to 40 million short tons – 10 percent of total coal production.[22] The message is clear: If we don't burn our coal at home, we will send it to consumers elsewhere, namely China (and increasingly India).

Whereas this exportation of coal might be helping the U.S. coal industry – as well as our carbon emissions – this approach is short-sighted. After all, pollution and emissions produced in China do not stay in China. This is especially disconcerting because Chinese power plants are not scrubbed for many pollutants and are inefficient, allowing them to emit more nitrous oxide, sulfur oxides, particulate matter and carbon dioxide per megawatt-hour of electricity that is generated than power plants in the U.S. In addition, by producing coal that is dedicated for export, we endure all the environmental impacts of producing it, but without the economic benefits of using it for power. Add in the transportation costs of shipping that coal across the Pacific, and it looks like the world is headed the wrong way on carbon.

Exporting Goods, Importing More Carbon Dioxide

The important factor to consider in this world transfer of carbon emissions is that a substantial fraction – about one-sixth or more – of China's coal consumption and resulting emissions comes from the manufacturing of products dedicated for the same countries that are trying to lessen

22 Over 12 months in 2021, the U.S. exported 85 million short tons; 15 percent of total coal production was exported, according to the U.S. Energy Information Administration.

their use of coal in the first place. It's like a massive parlor trick where we emit more carbon than ever before, but we do so through Chinese smokestacks instead of our own.

It works like this: Just as companies use offshore bank accounts to hide their cash flows, countries use offshore manufacturing to hide their carbon flows. As a consequence, the carbon emissions for European countries and the U.S. hold steady or decline despite economic and population growth, while China's emissions grow exponentially as the country accommodates its own growth while soaking up some of the emissions from developed countries.

This concept is referred to as "embedded carbon emissions." Researchers have determined that approximately 1 billion tons of carbon dioxide emissions in China are from the creation of products destined for export to other countries. You could say that carbon emissions are China's greatest import. Or they're our greatest export.

The irony is that American officials like to point to Chinese emissions as a reason not to engage in international climate negotiations. "After all, why on earth should we clean up our act if the world's dirtiest emitter doesn't do the same?" During those negotiations, we fail to recognize our own present and future complicity in shipping coal to China and in our demand for products from their factories. And there's a further complication: Climate change is a consequence of accumulated greenhouse gas emissions. Consequently, the emissions in any particular year aren't really the issue; it's the accumulation of emissions that is important.

On this tab, the U.S. stands far above the rest of the world. For example, although China emits more carbon dioxide on a yearly basis, the U.S. has emitted more carbon dioxide cumulatively in the last 270 years, by far: 417 billion metric tons by the U.S. and 236 billion metric tons by China from 1750 to 2020. In fact, the Chinese government often turns our carbon complaints against them back on us. They wonder why they should have to restrict their emissions in light of the fact that we emitted a lot of carbon for centuries while we grew our economy. All

China is doing is following in our footsteps. It's a fair argument, and I've never heard an effective counterargument.

Texas as a Proxy for China

The view from Texas is particularly interesting, because in many ways, Texas is the China of the United States – it's the dirty manufacturer with lax labor protections that makes things the nation wants (namely fuels, chemicals and concrete), but in the process, it emits a lot of carbon and consumes a lot of energy.

For example, Texans consume 60 percent more energy per person per year than the average American, who consumes three times as much energy as the average Chinese citizen. If Texas were a country, it would be the world's seventh-largest emitter of carbon dioxide. What happens when Chinese manufacturers and citizens consume energy like Texans? What if the rest of the world does the same? From where would we get that energy?

So, what does it mean for Texas and China that such a significant fraction of their emissions is the consequence of manufacturing goods that the rest of the world buys? Who owns those emissions? Texas and China? Or the people at the end of the supply chain who actually consume the products? Texas and China both have the same stakes in this quest. Setting up a way to attach a price to those emissions – whether through a carbon market or a carbon tax – accelerates reductions in emissions if that price can be passed on to consumers and if it's universally applied. If the U.S. has a carbon price, but not China, then we will just simply continue to offload our emissions onto their books, especially if they can't charge us for it.

Carbon regulations designed to punish carbon emitters will simply create an offshoring carousel as we push our carbon emitters from our smokestacks to others' and so forth. But if those prices can be passed to consumers through a carbon tax or some other scheme, then the ultimate consumers – the ones who in the end are an important driver of carbon emissions – can respond to the market signals by consuming less carbon-intensive products.

Ironically, in the face of massive job loss and concerns about competitiveness, it turns out a global carbon price might be our saving grace. Our factories and power plants are much more efficient than Chinese factories and power plants, so a price on carbon would help us gain a competitive advantage because our emissions per product would be relatively lower (and therefore cheaper). The same is true for Texas. If every refinery in the world had to pay a carbon price, Texas refineries – which are among the most efficient in the world, and therefore cheaper to operate in a carbon-constrained environment – would gain an advantage.

A Global Challenge

We Americans claim to be determined to reduce our energy consumption and to decarbonize society, but we've been going about it all wrong. Our solution has been to shift the consumption and emissions (and the economic gain) to China. And if that were not enough, we send them the coal they need to do it.

The proper way to regain our competitive advantage is to quit sending China our coal and our jobs and our money, and instead set up strict standards for the world's emitters, and then make sure we win the race in meeting them.

Conflict Between Russia and Georgia Adds New Twist to the Energy War

AUSTIN AMERICAN-STATESMAN, AUGUST 2008

How did it happen that Russia, which was economically and militarily crippled just a decade ago, can act as it wishes without risk of interference from the global community? How did the world lose its leverage? The answer: energy.

THE WORLD MIGHT NOT REALIZE IT, BUT RUSSIA'S ATTACKS ON Georgia are the latest example of a grand historical tradition – the energy war.[23]

The conflict has all the hallmarks of a typical energy war, which includes attacks on critical energy infrastructure. Newspapers reported that multinational energy companies are on alert as they watch the world's second-longest pipeline, which moves 1 million barrels per day across Georgia, suffer the onslaught of dozens of missiles from Russia. Whether the strikes targeted Georgia or its Western customers for the crude isn't clear.[24]

This episode also has a new element of a modern energy war: global complacence in the face of Russia's hydrocarbon dominance. As Russia flexes its muscles, the world stands aside. And that's what happens when the world becomes too dependent on a single country for its critical resources because of decades of inconsistent attention to our energy problem.[25]

23 Note that this piece was written in 2008. It was a similar story in 2022, but substitute "Ukraine" for the victim.

24 There were similar concerns in 2022 about the Nord Stream pipelines and other critical energy infrastructure.

25 Notably, in 2022 NATO stood unified in forcefully rejecting Russia's attacks, in contrast with its silence in 2008.

How did it happen that Russia, which was economically and militarily crippled just a decade ago, can act as it wishes without risk of interference from the global community? How did the world lose its leverage?

The answer: energy.

After the fall of the Soviet Union, Russia went through a rough transition. Built on the strength of its mineral extraction, it has since reemerged as a world leader. In 2008, Russia was the world's leading producer of petroleum (ahead of even Saudi Arabia) and the owner of the world's largest reserves of natural gas. These geological blessings combined with petulant nationalization of energy companies and seizure of multibillion-dollar energy projects have coincided with an era of record high prices for oil and gas.[26] Consequently, the Russian government's coffers are bursting, and its leaders overflow with patriotic confidence. These give them the ability – and the willingness – to issue muscular missives from their arsenal of petromilitarism.

This isn't to say the Georgia is blameless, but it's Russia's energy resources that allow it to be a more brazen aggressor. After all, the fossil fuel-powered missiles fly from Russia to Georgia, not the other way around.

Surprisingly, for once, it's not America's stupid energy policies that are the problem – though our energy policies are stupid. Rather, Europe's stupid energy policies are the problem. The EU, which is at greatest risk of the fallout from this conflict, is immobilized. Its silence is borne from a dependence on Russia for oil and gas that has disemboweled its ability to respond effectively.[27] Are Europeans really about to declare they want

26 Thanks partly to the Russian invasion of Ukraine in 2022, high prices returned.

27 This sheepishness in 2008 in light of Europe's dependence on Russian energy was a contributing factor to Russia's invasion of Ukraine in 2022.

to sit in the dark in their cold homes next winter to regain some leverage they lost to Russia? It doesn't appear so.[28]

Even though we aren't as directly affected as Europe, we can still help intervene. We can use our technological prowess to institute a high-profile crash program of oil and gas savings by swiftly and dramatically cutting our energy use with extensive conservation efforts, and quickly ramping up domestic production of oil, gas and alternative sources.

In doing so, we could dramatically lower global oil prices and consequently reduce Russia's revenues. We could then transfer our technologies to our allies in Europe to help them get out from under Russia's energy thumb.

And, while we're in the process of slowing Russia's aggressive postures by cutting its oil and gas revenues, we will enjoy the benefits of a mitigated environmental footprint from reduced consumption and the economic benefits of increased revenues for domestic energy technology companies.

Wouldn't it be great if our leaders in Washington recognized this conflict at face value – as a modern energy war? And unless we start taking drastic action for our energy policy at home and abroad, today's war will only be the first of many yet to come.

28 In 2022, it seems European attitudes had shifted, with aggressive pushes for conservation and alternatives to stand up to Russia's behavior.

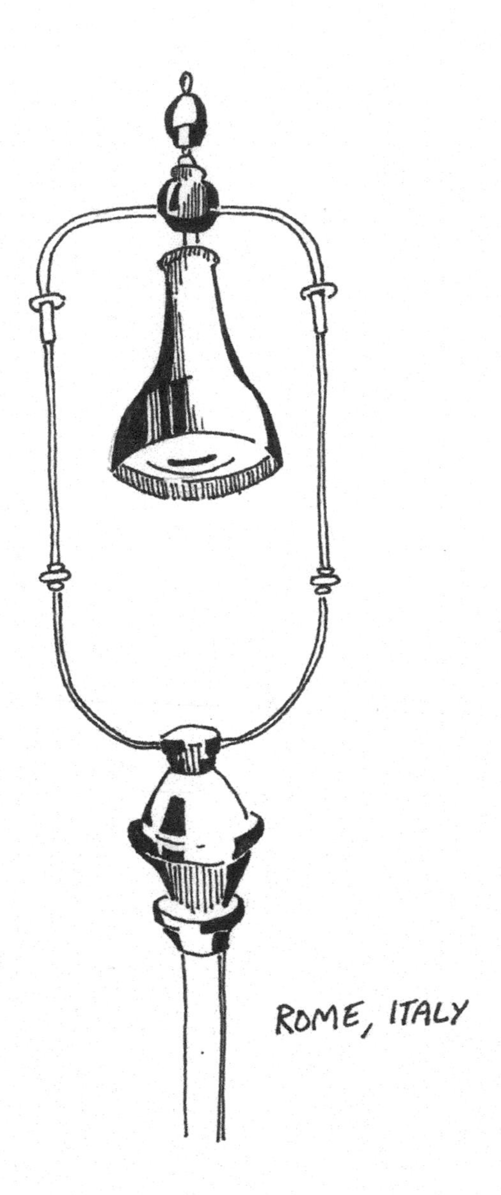
ROME, ITALY

Chapter 6

Connection Needs II: Energy & Transportation

Fly More, Not Less

MECHANICAL ENGINEERING, DECEMBER 2019

> *There's no shame in flying.*
> *But we must find ways to reduce aviation's climate impact.*

THE ENGLISH LANGUAGE HASN'T ADOPTED MANY WORDS FROM Swedish – smorgasbord, ombudsman come to mind – but another one that's been catching on is *flygskam*. It has been adopted by climate activists who have been using negative peer pressure, "flight shaming," to get others to give up long-distance travel or at least forgo travel by airplanes to reduce carbon dioxide emissions. The *New York Times* even ran a handy guide for how guilty we should feel when we fly.

Aviation is a convenient focus for climate activism. While the sector's contribution to global warming is small (around 2.5 percent of CO_2 emissions and somewhat more in total climate impact), it's growing faster than most other sectors. Aviation is also an activity that's experienced disproportionately by the wealthy. Nearly half of all Americans never fly in a given year, while two-thirds of the flights are taken by just 12 percent of the population.

The message this aviation inequality sends is that air travel is a playground for the rich – quite literally, the jet set – that requires leaving a scorched Earth behind as the price for an ephemeral contrail streaking across the sky.

I have a confession: I am a very frequent flyer. And I don't feel guilty.

Don't misunderstand me. Solving the climate crisis is absolutely humanity's most important challenge. But flying less doesn't solve it. If anything, we need to fly more, but do so in a low-carbon way.

I want to move to a brighter, more optimistic future by cleaning up our act rather than by depriving us of life's joys. And travel to faraway lands is one of those joys. In the middle of a rising wave of nativism, isolationism and nationalism, the last thing we should be doing is cut ourselves off from the common humanity of other cultures.

Travel can also be helpful to fight climate change. Renewable energy developers need to travel globally, and field researchers often must get to remote corners of the world where trains don't reach. Travel can be the glue that holds together international research collaborations, such as the multinational partnership that facilitated the lithium-ion battery's development, a feat that was recognized with the 2019 Nobel Prize in Chemistry.

Climate change requires global solutions and all hands on deck. Telling people they can no longer travel is not a great way to get them on board for tackling this crisis together.

Thankfully, there are several things we can do to reduce the footprint of aviation despite increased travel. For instance, we should reduce aviation's energy requirement, on a per passenger-mile basis, through such means as better managing airport congestion and ramping up airplane efficiency by use of lightweight materials, better wing design and improved engines. Electric-powered planes could be an option for particular routes, such as short-haul or nighttime flying (especially since they wouldn't be as restricted by noise pollution curfews).

Biofuels, which are already in development or use by the U.S. military and multiple commercial airlines, could be made the focus of a bigger R&D and testing push. Simultaneously, governments and industry could aggressively invest in synthetic jet fuel made from zero-carbon renewable electricity.

Airlines could also initiate a massive program of carbon offsets through direct carbon capture, soil management, afforestation, and so on, though the carbon offsets market to date has a spotty record. Airlines have a lot to lose from a movement that equates flying with sin, so some major airlines are already offering the opportunity to purchase offsets when buying a ticket.

Finally, a carbon tax on jet fuel would hit travel gluttons like me. It would also let people assess the value of every trip, rather than have choice taken away via a blanket prohibition on air travel.

As with other low-carbon pathways, Europe is taking the policy lead. The European Union's "Flightpath 2050" proposes the goal of reducing aviation's CO_2 emissions by 75 percent by 2050.

The United States should come up with similarly ambitious goals. We can do this. And we should start now with an accelerated program of R&D investments, tax incentives to encourage updating of airplane fleets, putting a price on CO_2 emissions to encourage low-carbon solutions, and better management of air traffic control globally.

Once we do that, we should take to the skies guilt-free – or perhaps *syndfri* – to see our beautiful world.

Michael E. Webber

A New Age of Rail

MECHANICAL ENGINEERING, FEBRUARY 2018

Reinvesting in freight railroads could be an infrastructure solution to multiple challenges.

EVERY DECADE OR SO, A FUTURISTIC TRANSPORTATION SCHEME captures the collective fancy. In the 1970s, it was magnetically levitated trains – maglevs – capable of shooting between cities at better than 300 miles per hour. In the 1990s, promoters of personal rapid transit were trying to drum up interest in automated minicars running on elevated tracks. Today, technologists breathlessly tout the Hyperloop, which would let vehicles zip through evacuated tubes faster than the speed of sound.

All those technologies sound cool and might eventually prove useful.

Today, however, we have some pressing transportation challenges. Our roads and bridges are poorly maintained and their limited capacity promotes traffic congestion. Transportation is the leading source of carbon dioxide emissions in the United States, and vehicle accidents kill more than 30,000 people every year.

Each of those problems are exacerbated by freight transportation on American highways. Freight is bigger than most people realize, moving about $55 billion worth of goods every single day in 2019. Trucks move 29 percent of the freight ton-miles, but are responsible for 77 percent of the sector's emissions. (Astonishingly, empty trucks account for about one-fifth of the truck miles traveled.) Between the rise of Walmart with its truck-based logistical system and the spread of Internet-based retailers such as Amazon, highway freight tonnage grew by 45 percent between 2000 and 2015 and hasn't stopped its upward trajectory.

According to the U.S. Department of Transportation, the existing population of trucks on congested highways already substantially impedes interstate commerce and projections suggest highway congestion will get much worse in the coming decades.

While the speed and flexibility of planes and cars has provided many benefits to society, the congested roads and airports signal they might be hitting their limits. Trucks are convenient because they enable flexible point-to-point operation, but they are relatively inefficient, dirty, dangerous and destructive to our roads.

These facts and trends reveal that freight – the movement of goods rather than people – presents a worthwhile opportunity for a system-wide improvement.

Rather than waiting for some still-unrealized technological breakthrough, we should instead expand our national freight rail system.

Yes, rail.

Road and Rail

It would be easy to believe that freight rail's day has passed. From its peak a century ago at more than 250,000 miles of total network mileage, today there are fewer than 95,000 miles of track for Class I railroads, as rail lost market share for the movement of people and goods to air travel and the Interstate Highway System. From 1990 to 2013 alone, the U.S. population increased 28.2 percent while track miles decreased 28.6 percent, despite increases in shipping and freight movement.

That decline didn't just happen. It was a policy choice carried out over decades. Starting in the 1950s we prioritized the movement of goods by truck over rail, investing trillions of dollars into the interstate and national highway system. Today, the national freight transportation infrastructure has about $6 trillion in assets, with more than half that total locked up in highways. While private sector trucks operate over public highways (making the Interstate Highway System effectively a subsidy by taxpayers to trucking companies), the freight railroads are almost entirely private.

Since bottoming out in the 1970s, financial restructuring enabled railroads to invest in better efficiency.

Consequently, revenues and incomes have risen for decades despite the lower trackage available. As one assessment by the RAND Corporation noted, American railroads have become extremely efficient and productive, moving increasing volumes of freight over a shrinking infrastructure.

In 2015, rail moved 40 percent of freight as measured in ton-miles, but is responsible for only 8 percent of freight transportation carbon emissions. Even though both trucks and locomotives use the same fuel – diesel – railways emit less CO_2 per ton-mile of freight movement because rail is much more energy efficient than trucking. By one estimate, moving freight by rail instead of trucks can save up to 1,000 gallons of fuel per carload.

Moreover, freight rail has the potential to get cleaner quicker than trucks, ships or planes. According to the Bureau of Transportation Statistics, the median age of the 25,000 locomotives in the United States in 2015 was less than 13 years, so natural fleet turnover patterns offer a chance to integrate newer, cleaner versions. That means the rail fleet can be cleaned through investments in just tens of thousands of locomotives (compared with more than 10 million heavy-duty trucks on the roads today). For instance, switching locomotives to compressed natural gas could reduce emissions while leveraging an abundant and secure source of fuel.

To be sure, trucking is great for the last few miles when delivering goods (which is useful, as most of us do not live next to train tracks). But more than two-thirds of the ton-miles of freight travel more than 500 miles and the efficiency of rail transportation means that even if the route is longer than what is possible with point-to-point trucking, shipping by rail would still use far less energy.

Reviving freight rail could also yield benefits beyond just energy savings and emissions reductions. Transferring freight from trucks to rail will also be safer.

Freight transportation is responsible for approximately 100,000 injuries and 4,500 fatalities each year, and trucks are responsible for 95 percent of the injuries and 88 percent of those deaths. Most of the people killed by trucks are in passenger vehicles sharing the road with semis. A study in 2013 concluded that the additional risk of fatalities from heavy trucks is equivalent to a gas tax of $0.97 per gallon.

By contrast, rail transportation is responsible for about 4,000 injuries and 500 fatalities, the vast preponderance of which are from trespassers on the railroad right-of-way. Increasing the amount of rail traffic might increase those numbers, but it would be more than offset by the reduction in deaths and injuries on the road.

Not only will roads be safer, but they will also be in better condition. The Highway Trust Fund that provides money for maintenance and repairs goes broke every year because revenues (collected via a tax on gasoline and diesel fuel) haven't kept up with expenses. One way to provide the trust fund with enough money is to raise the gasoline tax, set at 18.4 cents per gallon since 1993, but that is unpopular politically.

Another way is to reduce the wear and tear by removing the heaviest vehicles. Road damage scales with axle weight to the third power. So one 40-ton semi causes more than a thousand times the damage of a typical 4,000-pound car. As one seminal study noted, "For all practical purposes, structural damage to roads is caused by trucks and buses, not by cars."

Because of the governmental concern about the damage that heavy trucks can do to roads and bridges, significant effort and money is expended on vehicle weight enforcement. In 2013, there were just over 200 million weight enforcement activities for trucks, which identified about 400,000 violations, each of which puts our highway infrastructure at risk. By moving trucks off the road and onto rails, the rate of damage will decrease, reducing the risk of infrastructure failure and lowering maintenance costs.

Until recently, the freight railway network was being stressed on many routes by hauling of coal from fields in the Western U.S. to power plants around the country. Coal trains are huge: 100 hopper cars that can each carry 100 tons, all pulled by six 3,000-hp locomotives. In terms of ton-miles, coal still makes up the single largest commodity moved by the freight rail system.

Coal's decline, displaced by cheaper, cleaner natural gas, wind and solar power, opens up spare capacity in the rail system that could be used for moving other goods.

Those railroad rights of way from the Western coal fields to the East also have potential as routes for alternative energy. Following up on the old idea of lining train tracks with wires and poles, we could couple rail lines with a national high-voltage direct current transmission network, spanning the heart of the windy Great Plains and sunny Southwest, thereby enabling better integration of renewables, cleaning up the power sector further. Since the best solar and wind resources are often far from major demand centers like large cities, developing a national grid that can easily move power around would be advantageous.

We could even put those power lines underground to reduce their vulnerability to windstorms. Not only would a nationally coupled rail and powerline network reduce emissions and spawn more rural economic development, but it would also improve grid reliability while reducing the risk of sparking wildfires.

Laying electricity along the tracks also opens up the door for electrified freight trains. Such trains are common in Europe and it may be simpler to electrify freight rail transportation than to build out the charging infrastructure for electric road vehicles.

Switching Track

The vast preponderance of the rail system sprawls across the continent, so many of the economic benefits from a return to freight rail would accrue to rural areas. Since American companies like GE make locomotives[29], accelerating the adoption of newer, cleaner models would also trigger an uptick in domestic manufacturing jobs and output. In addition, locomotive engineers make about 30 percent more per hour than truck

29 GE sold its locomotive manufacturing unit to Wabtec in 2019.

drivers. Even the other jobs – rail yard engineers, signal and switch operators, conductors and so forth – earn more than truck drivers. Those higher wages would have rippling economic benefits.

Given all these potential cross-cutting benefits of increasing the role of rail for freight transportation, how should we proceed?

One simple way to encourage the switch from road to rail is to put a price on carbon. A carbon tax would harness the efficiency of markets while sending a price signal that rewards the more energy efficient and cleaner option of rail transportation.

Another approach – one that wouldn't also put motorists in the crosshairs – would be to raise money for road maintenance via a fee based on miles driven and vehicle weight. This would target the vehicles that do the most damage and stop the subsidy of heavy trucks by the drivers of small personal cars.

By more closely aligning the costs with the damage, trucking would lose some of its competitive advantage compared with rail.

While a carbon price and update to our highway tax model would likely encourage a lot of switching to rail for freight, increasing throughput (ton-miles) on rail without other improvements could degrade key performance metrics such as delivery time and reliability. Since many freight customers are very sensitive to those factors, commensurate investments have to be made in optimizing performance, double-tracking where possible, adding new tracks and alleviating bottlenecks.

Expanding track miles is an obvious step forward, though not the only one. Adding more sidings or double-tracking at congested zones can facilitate operation of more trains in different directions and allow trains operating at different speeds to more easily share the same track.

But just laying a bunch more track isn't enough.

As a pair of major studies by the RAND Corporation in 2008 and 2009 noted, increasing the national rail freight capacity will need a variety of strategies beyond direct infrastructure investments. Such measures include revised regulations, flexible pricing, deploying new technology and implementing improved operating practices. For instance, operational enhancements to more efficiently use existing tracks might be just as important as building more miles of track, but those changes need to be informed by more detailed and extensive modeling to identify locations of bottlenecks and developing schemes that avoid them.

Another way to increase capacity while cleaning up the transportation sector is to increase and improve the fleet of locomotives. Incentives for rail companies to buy newer, cleaner, more-efficient locomotives would simultaneously clean up and expand capacity.

More routine and detailed inspections of rail systems can also improve safety and throughput by allowing heavier freight loads and faster train movement. The Automated Track Inspection Program exists partly for this purpose. But according to one assessment, it "does not conduct a comprehensive evaluation of the national rail network on an annual basis due to the limited number of surveying cars." Let's invest in more surveying cars and conduct those inspections more consistently, frequently and comprehensively so that trains can move more efficiently.

New concepts will inevitably capture the imagination. We owe it to ourselves to investigate them to see if they will work. But in the end an old idea – moving goods by rail – might be the modern innovation we need to reduce energy consumption and avoid CO_2 emissions while making roads less congested, safer and more enjoyable for motorists.

Today's Decisions, Tomorrow's Cities

MECHANICAL ENGINEERING, OCTOBER 2019

Is building car-centric infrastructure a dead end?

TRANSPORTATION HAS EMERGED AS A STUMBLING BLOCK TO reducing carbon emissions. The energy density of liquid fossil fuels is a great match for powering vehicles. While supporting alternative fuel sources or electrification is a key step to reducing emissions, perhaps the best infrastructure solution is the one that can keep the vehicle parked: dense urban and suburban areas where destinations are close enough that you can walk, ride or scoot to them.

Unfortunately, building – or keeping – such places is easier said than done.

That's not because there isn't a demand. A recent survey by the National Association of Realtors found that nearly two-thirds of millennials prefer walking as their transportation for errands, and millennials prefer walking to work at twice the rate of baby boomers.

However, today's senior city leaders and transportation planners cut their teeth decades ago in an era of hollowed out inner cities and the rise of the suburbs predicated on single-family homes and privately owned automobiles. This movement to the suburbs created rush hours and traffic jams because of the coherence of people traveling in the same direction at the same time (from suburbs to cities in the morning and from the cities to the suburbs in the evening).

Today, many of those leaders and planners – scarred by the demands of peak-hour traffic, which still plagues them today – go back to the playbook in vogue when they were young: adding lanes to limited-access roads and highways at the expense of pedestrian connectivity. But widening highways to ease traffic congestion is a little bit like loosening our belts to solve obesity. Because our transportation infrastructure can last over a century, addressing today's problems with yesterday's solutions is an expensive proposition we will be stuck with for a long time.

Even in progressive, future-minded Austin, where I have lived longer than anywhere else and which experiences consistent growth over the decades, county and city planners are considering incredibly expensive concepts to create double-decker highways with flyovers that reach to the sky, disrupting views and bringing more noise, traffic and tailpipe pollution into the urban core. The city just finished an expensive, years-long, delay-ridden project to add a lane to one of the city's major highways – and traffic got worse after the project's completion.

While those lanes might help the out-of-towners passing through or the suburbanites commuting in for work, they are a blight that inequitably divides neighborhoods and worsens the quality of life for people who live in the city. That means planners are investing their limited funds to help people who live out of town at the expense of those who live in town, when they should be incentivizing the opposite.

Newly built automobile-oriented infrastructure may be popular with city fathers, but it's a market failure in the making. The *Wall Street Journal* picked up on the problem earlier this year while reporting on the real estate market. Baby boomers are having trouble selling their big houses in the suburbs because the next generation of home buyers want to live in walkable areas. This is the canary in the coal mine for a fundamental shift going on in front of our eyes, and this generational disconnect might mean we are doubling down on solutions that future urbanites reject.[30]

And, it's not just a generational disconnect, it's also a gender disconnect: The Realtors' survey found that women – especially young women–prioritize walkability and access to public transportation options more than men.

Some European cities that rebuilt after World War II on the model of car-centric American cities are now going back to their prewar roots. To reduce pollution and energy use, many of these cities are pushing to make the post-car urban landscape a reality by reducing speed limits, banning cars on certain days or entirely, and turning streets into pedestrian corridors lined with bike paths.

It's time for American cities to follow their lead. Instead of simply pouring more concrete, we should consider denser urban living with more options: mass transit, bike lanes, walkability – and a smaller carbon and

30 This article was written prior to the COVID-19 pandemic. Because of a pandemic-inspired push to work-from-home, suburban housing prices increased as many families sought more space. However, urban housing prices also increased, so there are many factors at work.

energy footprint. That way we are preparing for what the cities of the future want and need rather than what cities of the 1950s preferred.

The Electric Highway

MECHANICAL ENGINEERING, JUNE 2019

> *Electric and autonomous vehicles will change the layout and rhythm of our lives.*

THE DESIRE TO EXPLORE IS A DEFINING ASPECT OF HUMANITY. Transportation has shaped our societies: Where we can feasibly travel in the course of our day-to-day activities orders a great deal of our lives. And as our vehicles change, our concept of place changes along with it. Transportation has given shape to countries: Canals did it in the early 1800s; rail did it in the late 1800s through the first half of the 20th century, and highways did it in second half of the 20th century.

Electric and autonomous vehicles will change the layout and rhythm of our lives yet again. Entwined with this story of change are the fuels and forms of energy we used to enable the engines and motors to propel us forward.

Transportation can be powered by electric motors as well as by external (steam) and internal combustion engines. While the idea of electric vehicles seems like a modern concept, they have been around for more than a century and have intellectual roots that go back further. James Prescott Joule, the pioneer in thermodynamics, saw an electric future for transportation. "I can hardly doubt that electro-magnetism will ultimately be substituted for steam to propel machinery," he said in 1839. The transition to personal electric vehicles has been a long time coming.

Electric motors operate in a fundamentally different way than mechanical engines. They are inherently compact, quiet, and have fewer moving parts. They provide full torque even at low speeds, whereas mechanical engines give their highest power output at a few thousand revolutions per minute (rpm). That is why mechanical engines have complicated transmissions and clutches, so the driver can get a lot of power even when the car isn't moving or is at low speeds. Unfortunately, those additional moving parts, belts and crankshafts are all prone to failure, making the maintenance of mechanical vehicles more expensive.

Not only do electric vehicles have fewer moving parts that can fail, but also they do not produce fumes (because they do not have tailpipes) and they are much quieter. The gentle whir of an electric motor is much softer than the thousands of explosions per minute contained inside metal combustion engine blocks, which require muffling to comply with city noise ordinances.

Transportation systems that aren't continuously connected to electric rails or overhead wires (as trains and streetcars are) need to bring their own energy source with them. Such vehicles need an onboard storage device and a powerblock – a fuel tank and an internal combustion engine for a conventional car, or a battery and a motor for an electric vehicle. Hybrid vehicles have a fuel tank and a battery combined with a motor and an engine, which is one reason that hybrids sometimes cost more than their conventional counterparts. One challenge of this approach is that batteries are relatively pricey, heavy and bulky compared with gasoline. A tank of gasoline is a simple structure that holds a lot of energy. Getting 400-500 miles of range from a single tank of gasoline is a pretty standard achievement for modern hybrid cars, whereas it requires significant technical advance to get 200 or more miles of range from an electric car.[31]

Modern freight trains in the United States are diesel-electric trains. They carry diesel onboard, which is much more compact than a battery

31 My outlook was too pessimistic. In 2022 (three years after the article's publication), there were more than a dozen electric vehicle models available with a total range exceeding 300 miles, and at least one with a range exceeding 500 miles.

capable of pulling a train the same distance. The engine's sole purpose is to drive a generator to power an electric motor that drives the train's wheels. That way the train combines the energy storage benefits of diesel with the high torque and ease of control of an electric motor. In Europe and Asia, the train systems are highly electrified – and also much faster.

For buses or trash trucks, which are already heavy (and therefore would not be hobbled by the additional weight of a battery) and travel a fixed route before returning home at night, electricity is a compatible source because the batteries could charge while people are sleeping. In fact, the London Electrobus Company launched a fleet of 20 electric buses in 1907, and they worked fine for several years before the company shut down because of financial irregularities. A little over a century later, electric buses are making a comeback: Cities around the world are replacing entire fleets with electric buses as a way to reduce air pollution and reduce total cost of ownership.

A similar trend is afoot for personal automobiles. Some cities like Paris are banning diesel engines because of concerns about the air pollution emanating from tailpipes. Norway is offering steep incentives to consumers to support electric cars. In parallel, the cars are attractive to customers because of their quick acceleration and quiet operation. As a consequence of these converging factors, electric vehicle adoption is growing exponentially. This trend has economywide impacts because electrified drivetrains are more efficient than combustion-based systems.

Getting Cleaner

Transportation is a major energy user, responsible for about 28 percent of total U.S. energy consumption, mostly in the form of petroleum products burned in internal combustion engines operating with about 25 percent efficiency. That means that nine of the 12 gallons in a car's fuel tank are wasted (rejected as heat into the atmosphere) and only three are used for propulsion.

If we replaced all 3 trillion miles per year traveled by light-duty trucks and cars operating on 25 percent efficient combustion engines with 70 percent efficient electric vehicles, the economy's overall energy efficiency would be substantially improved and emissions would drop dramatically. If only wind, solar and nuclear energy were used for the electricity to charge those vehicles, then more than 1 billion metric tons (out of about 5 billion tons) of annual CO_2 emissions would be avoided. That means electrified transportation is not only a pathway to quieter, zippier operation, but is also cleaner and more efficient.

Expanding on that point, electric vehicles get cleaner over time as natural gas, wind and solar replace coal in the power sector, whereas combustion engines get dirtier with time as their systems degrade from normal wear and tear.

Electric transportation enabled one key innovation in mass transit: an extensive subway system. While the very first stretch of the London Underground system operated on steam locomotives that produced noxious fumes from the fuels they burned, smokeless electric trains were a much better fit for the poorly ventilated tunnels. With electrification, subway systems proliferated in the late 1800s through early 1900s. Subways transformed cities because they facilitated the mass movement of millions of riders without taking precious real estate or farmland on the surface. One observer noted that the subway essentially made New York City what it is, by bringing rich and poor people of many races and backgrounds together. With densification came a need for mass transit, which enabled more densification.

Compared with other transit systems, subways are a different kind of beast. Generally speaking, people don't want to live near airports because they are noisy, generate pollution and attract traffic jams. But people like to live near subway stations because the noise and fumes are out of sight and out of mind, and they offer great convenience for moving around a city. That's the result of electrification of mass transit, which together create a vast underground ballet of coordinated movement of people and machines.

Though the subway is well over a century old, it is a precursor to other concepts for underground transportation that moves people and goods at high speeds. I had the opportunity to work at the RAND

Corporation, the nation's oldest and most distinguished think tank, from 2004 to 2006. RAND was and is a special place where many good ideas flourish. RAND employees conceived of the communications satellite, the copay for health insurance, and the control deck for the starship Enterprise in *Star Trek*. I would occasionally thumb through their old reports because I was amazed at all the gems I would find.

One breakthrough report from 1972, "The Very High Speed Transit System" by Robert M. Salter, was particularly prescient. It laid out the concept of a high-speed, low-pollution alternative to air travel that used tunnels with high-powered pneumatic devices. It was inspired by the desire to save energy but also would avoid weather-related problems above ground. Imagine high-speed guinea pigs in their habit rails or the tubes at drive-through banks that use air to push or pull containers of your money from your car to the teller and back. That's the concept presented in the RAND report. It's also an ancestor of the core idea presented as Hyperloop by Elon Musk. It's as if there are no new ideas under the sun.

Robots on the Road

Turning to electric vehicles and switching from automobile ownership to ride-sharing in self-driving vehicles could reduce congestion by smartly controlling traffic flows using knowledge of where other travelers are going. Also, cars could be smaller. Most of us drive a single occupancy car that has room to seat four or five even though we rarely need the extra seats. With point-to-point mobility services for our commuting, vehicles could be tailored for the purposes of moving just a few bodies, which means they could be smaller.

Our recent research at the University of Texas at Austin has concluded that when the full life cycle costs of owning your own vehicle are considered (the cost of insurance and taxes on your garage at home, paying to park at work, maintenance, fuel and the lost productivity of time spent driving), using a mobility service like Uber or Lyft is the best economic option for more than a quarter of the population (using standard conditions from 2017). As the prices for mobility as a service drop, then that will be the economic option for a much larger fraction of society. Professionals who live in suburbs would benefit from using mobility services. Instead of wasting their time driving, commuters could rest, read emails, place phone calls or conduct other business. That work can create economic value – and reduce workers' office hours so they can get home earlier for dinner.

The rise of mobility services and self-driving cars might be as transformative as cars were in the first place. Imagine this scenario: We all have our own chauffeur who picks us up at our door the minute we're ready and drops us off at work or the grocery store, driving along roads with smooth traffic and sparing us the hassle and time of finding a parking spot. En route, we can read, text, think, sleep or talk on the phone without fear of causing an accident.

These mobility services with autonomous vehicles could save energy in a number of ways. Robotic drivers will be programmed to follow the best practices of driving – they won't have lead feet and bad habits. Embed more information into the cars and the surrounding infrastructure, and traffic will move more smoothly, reducing congestion, smog and energy consumption. A suite of connected cars that know what the other cars plan to do will make traffic lights obsolete; instead, the cars will continuously weave around each other at crossings or traffic circles. Safety will improve because each car will automatically know where the others are headed, reducing the risk of collision, just as planes do in the sky.

And since the cars will be better matched to the needs of the riders, there won't be lone commuters in gas-guzzling SUVs that only make sense on the occasional weekend. When you need to tow a boat or tote around your child's sports team and equipment, you could arrange for a robot-driven van or truck. Otherwise a smaller commuter car will be the primary vehicle of choice.

The cost of the vehicles will be shared through the mobility service company, keeping ownership costs down per mile traveled. Rather than each of us paying 100 percent for a $30,000 car we use 4 percent of the time, we will all pay for a more expensive car, but only when we need it, with one car meeting the needs of many. Auto insurance companies will likely have lower rates to reward the improved safety of robot-chauffeured cars compared to human-driven cars, creating a nice market incentive to get a ride.

People who really want to drive will pay extra insurance to reflect the additional risk they are introducing on the roads. Parents who really want their teenagers to drive can pay a premium yet higher. Once these technologies and market signals align to point in the same direction, the trends will be irreversible.

If you love to drive and worry that society will lose this critical skillset as we hand our transportation needs over to machines, then consider this: Some clever entrepreneur will sell you the opportunity to drive old beat-up cars in a circle on a dusty ranch while reliving the past. After all, that's what we do when we want to teach our kids how to ride horses.

Can We Escape Our Car-Centric World?

MECHANICAL ENGINEERING, APRIL 2019

> *Freedom has become just another word for nothing left to choose.*

THERE'S AN OLD JOKE ABOUT PEOPLE WHO DRIVE A MILE TO THE gym so they can walk two miles on a treadmill, but it is true that Americans typically don't travel from place to place on foot. That's not an indictment of laziness, however. The choice of getting around any way other than by petroleum-powered automobile has been taken away from most Americans.

To share an anecdote from my own personal experience: While my architect wife was remodeling the family home, we rented a house a few blocks from the local elementary school – so close that we could have enjoyed the daily ritual of walking together to the school. Unfortunately, walking simply was not an option. There wasn't a sidewalk along the busy road or a protected way to cross the streets safely. Sure, there was a crossing signal at the light, but no sidewalks on either side, so we would have had to walk in a ditch along a five-lane highway and stand in mud while waiting to cross.

Instead, we had to get in our car and drive those few blocks just for our own safety, consuming more energy and spewing more emissions along the way, only to wait in a long line of cars with other parents looking to drop off their kids, tailpipes close to kids' noses pumping particulate matter into their lungs. This isn't a good way to live.

Who's responsible for this mess?

It's easy to blame urban planners. By omitting protected bike lanes, sidewalks, pedestrian-friendly crossings, mass transit and density, their plans have essentially required us to use cars. But the dominance of one mode of transportation – petroleum-powered automobiles – above all others is also due to conscious decisions made by homebuilders.

As Jace Deloney noted via Twitter, modern houses are "designed like car-dominant cities. New homes often make the garage more prominent than the front door." When homebuilders erect subdivisions remote from public transit, they have decided for us that we will drive cars, since there are no alternative ways of getting around.

By not wiring or plumbing our garages for high-power charging or natural gas fueling, they have decided for us that we will drive gasoline or diesel-powered automobiles.

It's easy to imagine the alternative. If every garage in every home were built prewired and preplumbed for home charging and fueling, consumers would have more options about the type of car and fuel they use. Buying an electric or natural gas-powered vehicle wouldn't require hiring a contractor on top of choosing a car.

Ultimately, it's crazy that homebuilders choose our transport for us, when it's an easy fix with building codes and planning policies. Designing houses that accommodate different transportation modes – prewired and preplumbed for nonpetroleum cars and maybe with a bike rack too – is an important complement to designing livable, walkable cities that don't prioritize cars.

Our homes and buildings are the ultimate durable goods, lasting decades or centuries. And by one estimate, the world will need to add 2 billion homes over the next 80 years to account for population growth. Since new homes will be with us for a long time, building them with transportation and energy flexibility in mind will help us meet challenges we know about, such as climate change, and adapt to those that we haven't yet imagined.

To get there, Americans will have to reframe their notions of independence. The idea that unknown faraway bureaucrats or hidden corporatists would make decisions for us gets our hackles up and underpins the modern political rivalry between parties who strive to protect us from the oppression of either big government or big business. Self-reliance and the freedom to chart our own course is a definitive aspect of United States culture.

While the arrival of cars in post-WWII America typified freedom because they let us move to the suburbs and roam the country, we now have become trapped in car-dominated cities and homes. It's time we take the power back into our hands from homebuilders and city planners. We should demand more freedom, which means more options for movement.

This Pandemic Could Cause a Long-Term Shift in Car Ownership

MILKEN INSTITUTE, JULY 2020

The trend toward teleworking undermines some of the rationale for owning a car. When drivers only need cars for commuting two to three times per week, the economics of private car ownership worsen.

**Note: This was written in July 2020, just a few months into the COVID-19 lockdown. Updates have been included below as appropriate.*

THE CONVENTIONAL WISDOM IS THAT THE FEAR OF GETTING infected in crowded trains and buses will cause commuters to ditch mass transit to travel in private cars instead. Indeed, early data indicate increasing interest in car ownership.

But the long-term trend might actually be the opposite: The pandemic, together with more mobility options, might accelerate an irreversible decline in the traditional model of automobile ownership.

One of the most impactful and visible outcomes of the pandemic has been the rapid implementation of telework for wide swaths of the economy. In 2013, about 86 percent of U.S. employees drove to work in a private car. Another 10 percent of employees commuted to work by public transit, walking, biking or other means. The rest worked at home. By 2017, telecommuting had increased to just under 5 percent of the U.S. workforce. But that changed quickly in the pandemic: During the peak of shelter-in-place orders to minimize COVID-19 transmission, about half of employees did so.

How many of these will keep this practice for the long term?

Many managers who had resisted allowing their supervisees to work from home have been forced to get comfortable with the idea. In parallel, the tools for connectivity have vastly improved, making it more effective than before. Employers are discovering that their expensive leases, utility bills and tabs for office furniture can be dramatically reduced by encouraging employees to work from home at least a few days each week. Employees win by avoiding a time-consuming commute each day they work from home. Consequently, the Brookings Institution anticipates a permanent shift toward telecommuting.

Importantly, this trend undermines some of the rationale for owning a car. When drivers only need cars for commuting two to three times per week, the economics of private car ownership worsen. By contrast, for urban and suburban locations, mobility services and ride-hailing will be cheaper. Those services might be provided by cars (via Uber, Lyft, etc.) rather than mass transit, but they will compete with individual ownership. That means original equipment manufacturers will need to make cars designed for higher utility at a lower volume and get into services more aggressively.

Frankly, this trend was already afoot before COVID. A Wall Street Journal article in 2017 titled "The End of Car Ownership" laid out the case for ride-hailing to replace the century-old model of car ownership and noted that every major auto manufacturer was already experimenting with new business models to prepare for this shift.

My research at the University of Texas at Austin, together with the U.S. Department of Energy's Oak Ridge National Laboratory, demonstrated that as of 2017, it was already cost-effective for about a quarter of American drivers to use ride services instead of owning and driving their own car. A car's costs comprise much more than just the purchase price and gasoline – other costs such as the insurance, depreciation, parking at work and property taxes on the garage at home all add up. All told, the average cost of car ownership and operation is nearly $17,000 per year. The biggest line item in that tally is nearly $7,000 of lost productivity, which is the valuable time we spend behind the wheel of a car. With mobility services, that time is available for some other productive activity. A report released by Intel in June 2017 bills this newly useful time as the "Passenger Economy," and predicts it will be worth $7 trillion in 2050. Mobility companies will sell entertainment, business connectivity, meals-on-the-go and possibly even hair salon services.

So how do these things come together? The rise of mobility services means that commuters have more options to get to work than ever before. The rise of telework means that employees might commute to work half as often. If the economics of individual car ownership are already bad when cars are used 5 percent of the time, they will be even worse when they are used only 3 percent of the time.

These two factors combined mean that individual car ownership will face an inexorable decline as one of the lasting effects of the COVID pandemic.

How to Overhaul the Gas Tax

THE NEW YORK TIMES, DECEMBER 2013

> *The gas tax has remained stuck for three decades.*
> *It's time to consider modernizing it.*

WITH CONGRESSIONAL BUDGET NEGOTIATIONS PERPETUALLY under duress, and with our transportation infrastructure in many places crumbling before our eyes, it is time to consider modernizing the gas tax. Doing so would help fill the budget gap while updating and improving our transportation systems.

The Highway Trust Fund depends on federal fuel taxes for its finances. And those taxes have remained stuck for three[32] decades at 18.4 and 24.4 cents per gallon for gasoline and diesel, respectively. State taxes tack on another 31.1 cents per gallon on average.

Keep this fact in mind: There were about 260 million Americans in 1993 when the tax was last raised. Today there are more than 330 million. And we travel more miles than we did three decades ago. That

32 The article originally noted "two decades," but it has now been three decades since the federal taxes on gasoline and diesel were established at these levels.

means the transportation infrastructure has to do more with less per-mile spending, adjusted for inflation. That's why we see crumbling bridges on the news, outdated traffic-light patterns and clogged roads.

And, as we move into cities and use mass transit, we will drive less. As cars become more fuel efficient, they require less gasoline. At the same time, alternatively fueled cars such as electric vehicles don't pay gasoline taxes at all, and others, such as natural gas vehicles, pay a lower rate on average, so the current system subsidizes their use. That means our gasoline purchases – and our gas taxes – are declining, putting a strain on our trust fund.

The problem is already acute. Since 2000, spending by the highway fund has generally outpaced revenues; since 2008, the fund has required an infusion of $41 billion from the federal government's general fund. Since 2013, the typical mismatch in fuel taxes and expenditures exceeded $10 billion annually, a significant shortfall that Congress will be forced to make up.[33]

As Kim P. Cawley of the Congressional Budget Office argued in testimony before a House subcommittee, "The current trajectory of the Highway Trust Fund is unsustainable."

One choice, of course, would be to raise the gas tax. But this would surely raise the ire of many people and might be politically untenable. Some people have proposed a vehicle-miles-traveled tax, which could use fancy technologies like onboard GPS or mobile apps to track where we go and levy a fee in real time based on the distance, similar to electronic toll collection systems like E-ZPass. But these contraptions raise concerns about civil liberties and privacy protections (as the black box could report your whereabouts to authorities), and the extra technology might be costly to install. A blunt VMT tax also misses the fact that a light car does a lot less damage to the roads than a heavy truck. A better option is a "ton-mile" fee based on how

33 The Infrastructure Investment and Jobs Act of 2021 included some general revenue transfers to cover a portion of the projected shortfall out to 2031.

far vehicles travel and how heavy they are, so that all drivers pay their fair share to fix the resulting road damage. A one-ton car (which is typical for a compact car) that is driven 7,500 miles annually inflicts much less road damage than a two-ton truck that is driven 15,000 miles. While the gas tax captures some of that difference, as the truck driver would buy more fuel, it is not perfectly aligned.

Assessing a half-cent fee per ton-mile would cost a typical American car owner about $50 per year and would cover the highway fund's revenue shortfall, according to my calculations. And, rather than using tracking devices, the fee could be assessed during an annual sticker renewal or inspection that is conducted at state level: All the inspector has to do is read the odometer, look up the gross vehicle weight of the car's make and model, then assess the fee. With a fee on the order of two cents per ton-mile, gas and diesel taxes could be eliminated entirely.

So politicians could be credited for lowering the cost of gasoline at the pump while building a system that created fair competition between each fuel.

Switching to a ton-mile fee solves several problems at once: It raises the revenues we need for our transportation projects while ensuring that electric and natural gas vehicles don't get a free pass. It would also encourage people to drive both smaller cars and fewer miles, which would achieve additional benefits like reduced petroleum consumption, emissions, traffic congestion, and wear and tear on the roads and highways.

Republicans should like it because it would end the subsidies for alternatively fueled vehicles, and Democrats should like it because it would encourage energy conservation. In an era of partisan rancor, perhaps the bipartisan benefits of revising the gas tax would be one more step in the right direction.

Section IV

Esteem Needs

BRUGES, BELGIUM

Chapter 7

Esteem Needs I: Old vs. New Energy

What to Do About Natural Gas

SCIENTIFIC AMERICAN, APRIL 2021

> *If we can clean emissions out of the natural gas system, it could be part of a carbon-neutral future.*

IN THE MID-2010S IT BECAME COMMON TO SAY THAT NATURAL gas would be a bridge fuel to a zero-carbon future, in which solar, wind and other renewable technologies provide all of our energy without any carbon dioxide emissions to worsen climate change. But if natural gas is really a bridge, then it's not part of the long-term plan. And if we actually build the bridge, we're likely to stay on it.

Natural gas consumption in the U.S. has risen by a third in the past 15 years. Gas accounts for 32 percent of total energy consumption and is now the biggest source of electricity nationwide, largely displacing coal-fired power plants. Natural gas – primarily methane – burns much cleaner than coal does, and it provides ready backup to variable wind and solar farms. That sounds promising, except burning natural gas still creates CO_2. Methane in wells and pipelines can leak into the atmosphere, amplifying global warming. And once the last coal plant closes, natural gas plants become the dirtiest electricity sources.

To reduce CO_2 emissions, society has to decarbonize its energy systems as quickly as possible. Building more wind and solar farms is relatively inexpensive and fast, and it accelerates the shutdown of coal plants. But exploiting the best locations – the windswept plains and sunbaked deserts – requires a greatly expanded transmission grid to bring the electrons to major cities and manufacturing complexes. Those wires and poles introduce risks from windstorms, floods and fires – all rising because of climate change – and township after township routinely fights expansion plans: "Not in my backyard."

The natural gas infrastructure, almost all below ground, is far less prone to interruption. The U.S. has about 3 million miles of natural gas pipelines running underneath nearly every major city in the contiguous 48 states. After adding all the compressors, tanks and storage caverns, the infrastructure is worth several trillion dollars. The power plants themselves add hundreds of billions of dollars more. The nearly 70 million households served by natural gas have furnaces, water heaters and cooktops worth at least another $100 billion. Multiply all that sunk investment by about five for the entire world. Gas is also more intertwined than any other energy source with other sectors of society – transportation, buildings (for heating and cooking) and industry (for heat and as a feedstock for chemicals) – making it harder to replace.

Swapping out that infrastructure before its natural lifetime ends would also entail financial losses for the current owners, who will push back. The replacement technology could cost taxpayers, ratepayers and homeowners, who will push back too. And more electricity does not readily solve the need for liquid fuels burned in trucks, ships and planes or for intense heat in industrial foundries, distilleries and refineries that make volumes of metals, cement, glass, jet fuel and chemicals. The energy density of liquid fuels is difficult to match.

If we can clean emissions out of the natural gas system, it could be part of a carbon-neutral future instead of a bridge. The technology exists to extract the carbon or to transform the gas so that carbon coming out and carbon going in balance to zero or near zero.

The first step in a comprehensive plan for decarbonizing the nation's energy infrastructure would be improving energy efficiency and conservation to reduce consumption. The second would be to electrify as many cars, space heaters, water heaters and cook tops as is practical, using renewable sources. At the same time, tighten up the leaky gas infrastructure. And replace as much natural gas as possible with low-carbon alternatives such as biogas, hydrogen and synthesized methane or use a process called pyrolysis at the end of the natural gas pipes to get the carbon out.

Clean-energy supporters rightly worry that any investment in gas infrastructure creates a lock-in effect. Each new power plant, pipeline or gas storage unit has an expected lifetime of 25 to 80 years, so each element could either become a trap for more emissions or a stranded asset. But we can solve the lock-in problem with drop-in alternatives to natural gas: low-carbon gases that can flow through existing pipes, tanks and power plants, taking advantage of those trillions of dollars of assets.

Zero-Carbon Gas

The drop-in substitute most ready for natural gas is biomethane – methane gas produced from biological sources. Microbes inside large drums called anaerobic digesters chew up organic matter such as crop waste, manure, sewage and food waste and other garbage landfills, producing methane. Biodigesters, already a mature technology, transform waste streams at landfills and the waste lagoons adjacent to concentrated animal feeding operations from environmental liabilities into valuable commodities, generating revenues for municipalities and farmers.

Biomethane is working in Austin, Texas, Waste Management, which operates one of the city's landfills. The company collects biomethane from 128 wells on its site and burns it to generate enough electricity for 4,000 to 6,000 homes. And one of the city's wastewater treatment plants has eight biodigesters, each with 2 million gallons of capacity; microbes convert sewage into biogas that fuels on-site electricity generators.

About a quarter of the more than 2,000 U.S. landfills now harvest their gas or process their waste into biogas using biodigesters. That only offsets less than 1 percent of the country's total natural gas use, however. Biogas can serve as a direct substitute for natural gas, but the relative volume, globally, is low. If a farm, landfill or sewage plant cannot readily use the gas to make electricity or is not next to a gas grid, the biomethane might need to be liquefied and trucked to another location, reducing the carbon payoff. Still, biomethane is a commercially ready technology that can begin to decarbonize part of the gas system.

Hydrogen Instead of Methane

Natural gas can be replaced altogether, with hydrogen. Turbines can burn hydrogen to generate electricity for the grid, and internal combustion engines can burn it in heavy-duty vehicles. Hydrogen in fuel cells can produce electricity for cars, homes or offices. And hydrogen is a ready building block for many basic chemicals. Burning it, or reacting it in fuel cells, does not produce CO_2. Leaked hydrogen has a warming effect that is just a fraction of that of methane.

Natural hydrogen seeps out of the ground from basins in many cratons in the earth – large blocks of ancient rock that form the central parts of continents. Scientists have stumbled across these seeps for more than a century. Oil and gas companies, however, have considered hydrogen a nuisance when they find it alongside underground reservoirs because it can catch fire and can degrade metal piping. But today corporate and university researchers are drilling hydrogen test wells and launching multiyear programs to search for hydrogen underground. Anticipation feels similar to what arose during the very early days of fracking shale: a huge resource is out there, if engineers can figure out how to harness it cheaply and safely.

We can also manufacture hydrogen. Right now, most hydrogen for industry is produced from steam reforming of methane – adding heat and hot water to methane to create hydrogen and CO_2. Electrolysis

– using electricity to split water into hydrogen and oxygen – can also create hydrogen gas. Both processes require significant amounts of energy, however.

Moving and storing gaseous hydrogen is also a challenge. Because of hydrogen's low density, it takes a lot of energy to move it through a pipe compared with denser gases such as methane or liquids such as petroleum. After several hundred kilometers, the inefficiency makes moving hydrogen more expensive than the value of the energy it carries. And hydrogen can embrittle steel pipelines unless that is mitigated by altering operating conditions or incorporating expensive alloys.

One way to integrate hydrogen is to mix it with methane in an existing natural gas pipeline. This blending decarbonizes some of the system by displacing a portion of the natural gas with hydrogen. Experiments in the U.K. and France show that a mixture of 80 percent methane and 20 percent hydrogen can be efficiently moved in a natural gas pipeline. As part of a study from mid-2018 to March 2020, Dunkirk, France, used an 80-20 blend of methane and hydrogen to fuel 100 homes and a hospital boiler without any new equipment along the pipeline or in the buildings.

Fittings inside furnaces and stoves, such as burner tips, might need to be altered or replaced for blends with more than 20 percent hydrogen because, like pure hydrogen, blended gas burns at different temperatures and rates.

Common hydrogen carriers such as ammonia (NH_3), formic acid (CH_2O_2) and methanol (CH_3OH) are liquid at near-ambient conditions, making them easier to transport than gaseous hydrogen. Although ammonia is caustic, it is already moved worldwide as a fertilizer ingredient, and it can be burned without producing any CO_2. Methane (CH_4) could be the most efficient option because it carries four hydrogen atoms for every carbon atom and is already compatible with existing pipes, compressors, tanks, turbines and appliances.

Demonstration projects are growing quickly in number. Finnish industrial builder Wärtsilä is constructing a new ship for 2023 named Viking Energy that will run on ammonia with fuel cells, avoiding greenhouse gas emissions and other pollutants that plague the maritime sector. Air France and the Charles De Gaulle airport in Paris are very interested in hydrogen as a way to decarbonize aviation. Hydrogen carriers are still in the early stages of research, however, so it is difficult to say how successful they could be.

Power plants that burn hydrogen are on drawing boards too. In Delta, Utah, the Intermountain Power Plant – one of the largest U.S. coal-fired plants – sends electricity hundreds of miles to Los Angeles. To meet the city's long-term requirement for renewable and low-carbon energy, in 2025, plant owners will replace the coal boilers with turbines that can burn hydrogen. They will start with a blend of 30 percent hydrogen in natural gas and will shift to 100 percent hydrogen later. The hydrogen will be generated right there using electrolysis powered by wind and solar and will be stored in more than 100 existing underground salt caverns, each about the size of the Empire State Building.

End of the Pipe

Instead of decarbonizing natural gas before it goes into the pipeline, we could remove the carbon at the end of the pipe, where customers consume the gas. Methane, for example, can be split at the user's location into hydrogen and solid carbon, which looks like a fine, black dust. The process, called methane pyrolysis, is efficient and eliminates CO_2 emissions. Every kilogram of hydrogen produced from pyrolyzed methane generates three kilograms of solid carbon instead of nine kilograms of CO_2 gas that would be emitted if the methane was burned.

The pile of carbon dust that accumulates inside a collector in a furnace or stove would be carted away each month or so. The carbon piles actually have value, though, because they can be sold as a basic ingredient for making graphite, rubber, coatings, batteries and chemicals, as well as a soil amendment for agriculture.

Although engineers have studied methane pyrolysis for decades, they have deployed it only in small demonstration projects. Some equipment at the end of the pipe has to be changed to separate the carbon, but no expensive hydrogen pipelines would have to be built, simplifying matters greatly. Pyrolysis of conventional natural gas can bring the entire system to nearly zero carbon. Adding methane from biodigesters or made from CO_2 in the atmosphere using renewable electricity could make the system carbon negative.

Imagining any of these decarbonized futures might conjure up visions of large new industrial complexes or millions of small equipment changes for consumers. But so do other proposals to curb emissions. Electrifying every heater, stove and vehicle would require widespread technology replacement. Plans to directly pull CO_2 from the air would require millions of big machines to capture the gas and sequester it – sprawling enterprises that would also demand lots of new land and new electricity.

Decarbonized gas would let us take advantage of trillions of dollars of existing pipelines, equipment and appliances, saving huge sums of money and years of time in creating a zero-carbon energy system. We would, of course, have to fix the leaky infrastructure. Leaks can be minimized by replacing pneumatic equipment with electric devices at well sites, improving the automation of pipe and tank inspections with sensors on drones and robots, and writing regulations that no longer turn a blind eye to leaking, deliberate venting or burning unwanted gas. This would create jobs for workers in the oil and gas industries and would clean up the energy infrastructure, which in turn could lessen pollution in communities near energy facilities.

Reining in climate change requires many solutions. Declaring who cannot be part of those, such as natural gas companies, only raises resistance to progress. Because decarbonized gas can complement renewable electricity and because it might be a faster, cheaper and more effective path for parts of society that are difficult to electrify, we should not discard gas as an option. We have a massive gas infrastructure, and we have to figure out what to do with it. Scrapping it would be slow, expensive and incredibly difficult, but we could instead put it to work to help create a low-carbon future.

It's Complicated

MECHANICAL ENGINEERING, JUNE 2019

It's time we figure out our relationship with nuclear energy.

IMAGINE WE HAD STATUS BUTTONS ON OUR ENERGY SYSTEM'S dashboard the way social media users reveal their personal connections. Decades ago, we might have selected something like "happily married" for coal, hydro and nuclear power. All were steady or growing in importance, and society was committed for the long term.

Today, our relationship with energy is messy. Millennials would say we're in the process of ghosting coal. Wind, solar and natural gas are the new BFFs – best friends forever.

And nuclear? As a Millennial might put it: "It's complicated." We can't decide if we're ready for a break or a long-term commitment. And that confuses the situation for other would-be suitors.

This lack of a clear signal matters because we need to simultaneously increase energy access in the developing world – where around a billion people still don't have access to electricity – while decreasing energy's climate impacts in the developed world. The dual challenge is daunting and knowing whether nuclear is or isn't part of our long-term plans is critical.

The problem is that there's no single correct way to think about nuclear power.

I am engineer, so I look at nuclear power as a clean, safe, reliable source of low-carbon baseload power. In that lens, I'm ready for a serious relationship.

But lawyers look at nuclear as a liability waiting to happen. And businessmen in the natural gas, wind and solar industries see it as a competitor.

Many grid experts like nuclear because it has high reliability and is hardened such that it can work even through a hurricane. Other grid experts worry about nuclear because it ramps slowly and is prone to shutdowns during heat waves because the cooling water isn't sufficiently available or cold enough to guarantee the safety of the plant.

Financiers and energy economists see nuclear as an expensive source of power because it costs so much to build. Cost overruns are a typical part of the story for new power plants, and the capital sunk into one plant could buy several times more capacity from other sources.

Yet other energy economists look at nuclear as an affordable option because it features stable output with low and predictable operating costs, in contrast with the volatility of natural gas prices and the variable nature of renewables.

Some national security observers worry that spreading nuclear power throughout the developing world will worsen weapons proliferation. Others worry that without an active nuclear power fleet in the United States, weapons proliferation will worsen because we won't have the experts trained to sniff out rogue programs abroad.[34]

Environmentalists? Sure, some look at nuclear power as a source of long-lived radioactive solid waste we inject into Earth's crust. But others see nuclear as a source of emissions-free energy that avoids the long-lived carbon waste that we inject into the atmosphere. Public health experts, meanwhile, tout the fewer asthma epidemics and ozone action days that come with nuclear, while at the same time worrying about the risks of meltdowns and radioactive leaks that nuclear poses.

In total, these conflicted feelings are why new nuclear plants have a hard time being supported, financed and built. It's just easier to build new natural gas combined-cycle power plants along with wind and solar farms. Ultimately, we like to do the easy things first.

From my perspective as an engineer working in France, where nuclear provides more than 70 percent of the electricity consumed in

34 For more on that point, please see "Why the withering nuclear power industry threatens US national security," by Michael E. Webber, *The Conversation,* Aug. 11, 2017. Source: https://theconversation.com/why-the-withering-nuclear-power-industry-threatens-us-national-security-82351 [Accessed Nov. 15, 2022]

the nation, I see how a long-term commitment can work. I think it's time we invest in R&D programs and public policies that incentivize smaller, cheaper, cleaner and faster nuclear power plants that we can build quickly in line with our cost expectations while helping us improve electricity access and reduce pollution.

In other words, I'm looking for the perfect soulmate. Which rarely exists in the real world.

Renaissance and Revolution

MECHANICAL ENGINEERING, APRIL 2018

Nuclear power's long-delayed revival is a victim of the rise of shale gas.

GLOBAL POWER MARKETS ARE IN FLUX, REACTING TO SUCH FORCES as the rise of renewables, smarter grids, worries about cyberattacks, needs for resilience in the face of more extreme weather, and energy sector decarbonization. Missing from the headlines, unfortunately, is the fate of nuclear power.

Nuclear energy is a large source of carbon-free power, but it's being battered from multiple directions. Variable renewables like wind and solar are making the grid more dynamic on top of already wide swings in demand due to weather. That means nuclear is being asked to be something it was never designed to be: flexible, nimble and fast. In some places, nuclear power plants are being shut down in favor of fossil-fueled plants – a step backward from an emissions perspective.

As nuclear struggles to find its place in competitive markets, we might look to the shale revolution for answers.

The rapid rise in economical extraction of oil and gas from shale formations is the most important story in energy geopolitics from the last decade. In the United States, this development has shifted the conversation from how to finance ever-growing imports of crude oil and liquefied natural gas to debates weighing the opportunities and disadvantages of becoming a net energy exporter.

The shale revolution is the outcome of three converging factors. First, a decades-long partnership between the U.S. Department of Energy and shale pioneer George Mitchell provided a stable policy environment that allowed the freedom of experimentation to find a solution that would work. In addition, the Energy Policy Act of 2005 affirmed that hydraulic fracturing would not be regulated the same as wastewater injection, which opened up the floodgates of capital from investors who had been wary of regulatory and permitting authority uncertainty.

The second factor was technological. It took a novel combination of advanced technologies – primarily horizontal drilling and hydraulic fracturing – to make production from shale formations economical.

Those two aspects – policy stability and technological advancements – were critical but not sufficient. Market forces provided the third ingredient. Soaring natural gas prices in the first decade of the 21st century along with growing demand for energy sent a market signal to producers to increase supply. And so they turned to shale.

Could we replicate that playbook for nuclear power?

U.S. nuclear policy has been supportive for decades. A significant fraction of the U.S. Department of Energy's annual R&D budget goes to nuclear energy, the government provides tens of billions of dollars in loan guarantees, and the Price-Anderson Act of 1957 reduces the financial risk to operators from accidents at nuclear power plants.

The nuclear industry is also enjoying a technical revolution: The advent of small modular reactors, with their self-contained design, dry cooling (to avoid water requirements), and smaller size simplifies permitting and reduces the total system cost.

If nuclear is following the shale playbook by integrating pronuclear policies with nuclear technology advancement, why hasn't a nuclear renaissance begun? The answer is because it's missing the third element: market signals that prompt stakeholders to act. Instead, the market signals – composed primarily of stubbornly low electricity prices – are pointed in the wrong direction.[35]

While the shale revolution lays out the blueprint for success, it is also the reason why the nuclear revolution hasn't yet begun. The exploitation of shale formations has yielded an ocean of cheap natural gas, which has depressed the prices that electric generating assets can charge.

Nuclear power plants will struggle until we send robust, pronuclear market signals, such as increasing natural gas prices or putting a price on carbon, which would make nuclear power's competitors more expensive. Until the markets can align with the technologies and policies, nuclear power will just sputter along. The stakes are high, so it's time we figured this out.

Making Renewables Work

MECHANICAL ENGINEERING, DECEMBER 2016

> *Grid operators face real challenges in integrating wind and solar power. Maybe we need to rethink the relationship between electricity supply and demand.*

AS THE U.S. PUSHES FOR CLEANER FORMS OF ENERGY, SOLAR AND wind power are increasingly popular choices. To the consternation of grid operators pushing for high reliability, however, wind and solar power plants do not behave like traditional power plants. Rather than being dispatched by a central control center, wind and solar resources come and go according to a schedule set by the forces of nature.

35 In 2022, electricity prices were much higher than in 2018 and indeed nuclear seems to be gaining traction again, with closures delayed, new plants coming online, and new licenses for small modular reactors.

Conventional wisdom holds the very clear and firm conviction that this makes renewables bad for reliability. Many esteemed organizations and deep thinkers have declared that once renewables pass a fairly low percentage of grid capacity, usually something like 20 percent, the system faces severe challenges.

But is that true?

Unquestionably, wind and solar are variable and that variability can be problematic. The sun doesn't shine when it is cloudy or nighttime, and wind is far from steady. If the wind dies down suddenly or if we have weeks of overcast skies, then the balance of the grid must fill in the gap, which can be expensive or a strain on the entire system.

Wind is particularly vexing because in the midcontinent, where we have abundant wind resources, the wind blows strongest in the spring in the middle of the night – which is when demand for electricity in North America is the lowest. And onshore wind is weakest on the hot summer afternoons when we need it the most.

Solar power matches up a little better. It is most abundant on sunny days, when our demand is usually highest. But solar energy has its own problems: Sunshine peaks in the early afternoon, but the latency in heating up the thermal masses such as the air, land and water shifts the hottest part of the day to a few hours later, as the sun moves toward the horizon. That means just as demand is rising from air conditioners and people coming home from work and turning on their residential appliances, the amount of electricity produced from solar panels is crashing and dispatchers have to rapidly ramp up many other thermal generators to balance the load.

These ramps can be as steep as several gigawatts of capacity coming on line each hour. Thermal plants don't like to be cycled up and down, so each of these rapid ramps induces strain on the equipment, driving up costs and introducing safety risks.

In sunny California, which has the highest penetration of solar panels in the nation, the gap between total load and the part that solar doesn't satisfy is called a duck curve – a fat middle during the heart of the day followed by a steep ramp upward, which is the ducks' neck. In

Texas we call it the dead armadillo curve, as it is reminiscent of the familiar roadkill.

Are those symbols of what renewables do to grid reliability? They don't have to be.

Turning power plants up and down so that the supply matches our demand represents legacy thinking. Instead, what if we turned our demand up and down to match when the supply is available? Electricity storage systems – such as batteries, pumped hydropower or compressed air in underground caverns – can help us achieve that goal, storing wind and solar electricity when it's available and letting us consume it later when we need it or want it.

We can also shift a lot of our processes to work sometime other than the peak of the day. Treating water, operating data centers and many types of manufacturing can be done flexibly. Pumps for municipal water systems often work 8-to-5, coinciding with the worker's shift, but a lot of those processes could be done just as easily at night. There are many more examples of this type of load-shifting. In other words, we should ditch the load-following mindset (operating power plants when demand is high) for a more modern supply-following mindset (operating factories when supply is high).

It will take some ingenuity and an embrace of a new relationship between electricity supply and demand, but a renewable-dominated power system can be just as reliable – or even more reliable – than what we have today.

Are Solar and Wind Really Killing Coal, Nuclear and Grid Reliability?

THE CONVERSATION, MAY 2017

By Michael E. Webber, Joshua D. Rhodes, Thomas Deetjen and F. Todd Davidson

> *Incorporating wind and solar into the grid along with fast-ramping natural gas, smart market designs and integrated load control systems will lead to a cleaner, cheaper, more reliable grid.*

U.S. SECRETARY OF ENERGY RICK PERRY IN APRIL 2017 REQUESTED a study to assess the effect of renewable energy policies on nuclear and coal-fired power plants.

Some energy analysts responded with confusion, as the subject has been extensively studied by grid operators and the Department of Energy's own national labs. Others were more critical, saying the intent of the review is to favor the use of nuclear and coal over renewable sources.

So, are wind and solar killing coal and nuclear? Yes, but not by themselves and not for the reasons most people think. Are wind and solar killing grid reliability? No, not where the grid's technology and regulations have been modernized. In those places, overall grid operation has improved, not worsened.

To understand why, we need to trace the path of electrons from the wall socket back to power generators and examine the markets and policies that dictate that flow. As energy scholars based in Texas – the national leader in wind – we've seen these dynamics play out over the past decade, including when Perry was governor.

Wrong Question

There has been a lot of ink spilled on why coal is in trouble. A quick recap: Natural gas is plentiful and cheap. Our coal fleet is old and depreciated. Energy use in the U.S. has flatlined, so there's less financial incentive to build big new power plants.

Part of Perry's review is aimed at establishing how wind and solar, which are variable sources of power, are affecting so-called baseload sources – the power plants that provide the steady flow of electricity needed to meet the minimum demand.

Posing the question whether wind and solar are killing baseload generators, including coal plants, reveals an antiquated mindset about power markets that hasn't been relevant in many places for at least a decade. It would be similar to asking in the late 1990s whether email was killing fax machines and snail mail. The answer would have been an unequivocal "yes" followed by cheers of "hallelujah" and "it's about time" because both had bumped into the limits of their utility. How quickly 1990s consumers leaped to something faster, less impactful and cheaper than the older approach was a sign that they were ready for it.

Something similar is happening in today's power markets, as customers again choose faster, less impactful, cheaper options – namely wind, solar and natural gas plants that quickly boost or cut their output – as opposed to clinging to the outdated, lumbering options developed decades before. Even the Department of Energy's own analysis states that "many of the old paradigms that govern the (electricity) sector are also evolving."

Wind and solar are making older generators less viable because their low, stable prices and emissions-free operation are desirable. And they aren't hurting grid reliability the way critics had assumed because other innovations have happened simultaneously.

Texas Pioneer

Let's use the case study of Texas to illustrate. Since Texas has its own grid, known as the Electricity Reliability Council of Texas or ERCOT, and has installed more wind capacity than the next three wind-leading states combined, the Texas experience shows what variable renewables like wind power do to the grid.

In competitive markets like ERCOT, companies that run power plants place bids into an auction to provide electricity at a certain time for a certain price. A bid stack is jargon for the collection of all these bids lined up in order by price in auction-based markets (such as Texas).

Markets use bid stacks to make sure that the lowest-cost power plants are dispatched first and the most expensive power plants are dispatched last. This market-based system is designed to deliver the lowest-cost electricity to consumers while also keeping power plant owners from operating at a loss.

Throughout the day, the market price for electricity (in $/ Megawatt-hour) changes as demand changes.

The cost of natural gas also affects the price of electricity. As the price of natural gas drops, each of the natural gas power plants drop in price. That's no surprise: When it costs less for them to operate, they can bid a lower price into the market and move earlier in the line.

When gas drops in price, it begins to displace coal as a less-expensive source of electricity. This scenario reflects today's reality: Gas is relatively cheap, so grids are using it for more of our electricity than coal.

How do renewables affect the bid stack? Renewable sources such as wind, solar and hydro have no fuel costs – sunlight, wind and flowing water are free. That means their marginal operational cost is near zero; the cost is essentially the same to operate one megawatt of wind as compared to the cost of operating 10 megawatts of wind since generators don't need to buy fuel. That means as more wind and solar farms are installed, more capacity is inserted at the cheapest end of the bid stack.

This insertion pushes out other generators such as nuclear, natural gas and coal, causing some of them to no longer be dispatched into the grid – that is, they don't supply power into the grid (or get paid). So as more renewables are installed, power markets dispatch fewer conventional options. And, because the marginal cost of these new sources is almost free, they substantially lower the cost for electricity. This is great news for consumers (all of us) as our bills decrease, but bad news for competitors (such as coal plant owners) who operate their plants less often and are paid less when the plants do operate.

What does all this mean? Natural gas and renewables are affecting coal in two ways. Natural gas is a direct competitor with coal because both can be dispatched – turned on – when a grid operator needs more power. That is helpful for grid reliability. But, as the cost of natural gas has fallen, coal has become less competitive because it is cheaper to operate a natural gas power plant.

The effect of renewables is slightly different: Wind and solar power are not dispatchable, so they cannot be turned on at a moment's notice. But, when they do turn on, during windy evenings or sunny days in Texas, they operate at very low marginal cost and thus operate very competitively.

Research at the University of Texas Austin shows that while installing significant amounts of solar power would increase annual grid management costs by $10 million in ERCOT, it would reduce annual wholesale electricity costs by $900 million. The result of all this is that renewables compete with conventional sources of power, but they do not displace nearly as much coal as cheap natural gas. In fact, cheap gas displaces, on average, more than twice as much coal than renewables have in ERCOT.

What About Nuclear?

Nuclear's problems are largely self-inflicted. In short: The price to build nuclear is high, so we don't build many nuclear plants these days. Since we don't build, we don't have the manufacturing capability. Since we don't have the manufacturing capability, the price to build nuclear is high. Since the price to build nuclear is high, we don't build nuclear these days ... and so on and so forth.

Today, cheap gas, having already beaten up on coal, is a threat to new nuclear power plants and less-efficient older plants. New natural gas combined-cycle power plants can be built for about one-sixth the cost of a new nuclear plant, are almost twice as efficient and you can build them in smaller increments, making them easier to finance.

Market Innovation and Technology Can Fix Reliability

Because wind energy comes and goes with the weather, it makes grid operators nervous. But wind forecasting has improved dramatically, giving more confidence to those who need to keep the lights on.

And, interestingly enough, the requirements for reserve capacity (backup power for when wind power dips) to manage the grid smoothly went down, not up, over the past few years in Texas, despite rapid growth in wind during Governor Perry's tenure. That is, the costs for managing variability in the grid decreased.

Why has there been little disruption to the reliability of the Texas grid?[36] Because alongside rapid growth in wind installations was a market transformation in ERCOT. While Secretary Perry was governor, the Texas market went from a coarse, slow market to a fine-tuned, fast market. Innovating the market to one that is dynamic and fully functioning made it easy to include more wind into the system. It's also a sign of how advanced technologies enable us to reinvent the grid toward one that is cheaper, cleaner and more reliable.

But there is still more to do – information technology coupled with integrated hardware can help. Consider this: There are 7.7 million smart meters in Texas, most of them residential. We've estimated that installing 7 million controllable thermostats for just the households in Texas would cost $2 billion. Residential air conditioning is responsible for about 50 percent of peak demand in Texas in the summer. That means about 30 gigawatts of peak demand in Texas is just from residential air conditioners.

By dynamically managing our air conditioning loads – that is, adjusting thermostats to lower overall demand without impacting people's comfort – we could reduce peak demand by 10 to 15 gigawatts. That

---

36 The Texas grid famously suffered significant grid outages in February 2021 during a winter storm. According to official after-action reports by the Public Utility Commission of Texas and the Federal Energy Regulatory Commission, outages by traditional thermal power plants (gas, coal and nuclear) were the biggest problems.

means we might not need $10 billion to $15 billion worth of power plants. Spending $2 billion to avoid $15 billion is a good deal for consumers. In fact, you could give the thermostat away for free and pay each household $700 for their trouble and it would still be cheaper than any power plant we can build.

In the end, Secretary Perry has posed good questions. Thankfully, because of lessons learned while he was governor of Texas, we already have answers: Despite concerns to the contrary, incorporating wind and solar into the grid along with fast-ramping natural gas, smart market designs and integrated load control systems will lead to a cleaner, cheaper, more reliable grid.

How Dependable Is the Traditional Grid?

MECHANICAL ENGINEERING, OCTOBER 2016

Utility managers worry that renewables may hurt grid reliability. In fact, all energy forms and technologies suffer from some sort of reliability challenge.

ONE OF AESOP'S FABLES TELLS OF THE HEARTY OAK TREE THAT mocks the reeds for weakly bending with the wind. The strong oak resists the wind for many decades, but one day it snaps in a windstorm, falling apart in spectacular fashion. The reeds, meanwhile, live on, flexible and responsive to change. The moral of the story is that it is better to bend, not break.

Many observers of our electric utilities mock renewables for their weakness, in the form of their variability and intermittency, and confidently praise large baseload power plants for their steady reliability. I believe that confidence is misplaced. I want to challenge the idea that only renewables suffer from reliability issues.

In fact, all energy forms and technologies suffer from some sort of reliability challenge.

Coal suffers from distribution problems. One train wreck or bad section of track 1,000 miles away can delay a coal train for weeks. When a polar vortex froze over lakes and rivers in 2014, the barges delivering coal couldn't get through the ice, threatening a coal-fired power plant along Lake Superior with a fuel supply disruption. Coal-carrying barges also have trouble delivering fuel along the Mississippi River when major droughts strike.

These factors are why coal plant operators store 60 days' worth of coal on-site. They've built the likelihood for disruption into their operational schemes.

Of course, foul weather can cause other problems too. A major cold snap in 2011 in Texas caused some water pipes at coal plants to freeze, forcing the plants off and triggering statewide blackouts.

Other forms of baseload have reliability issues.

Historically, the biggest concern for grid operators was the risk that a nuclear power plant would trip offline, erasing 1 gigawatt or more of capacity in a second or less. In Texas, the reserve margin – the amount of power from backup plants that needs to be on and ready to generate electricity on a moment's notice – is set based on the risk that two nuclear reactors, or 2.6 gigawatts of generating capacity, would go offline simultaneously. That grid operators have baked nuclear variability into their planning is very telling.

Natural gas can suffer from low pressure in the pipelines when demand is unusually high for home heating and power plants simultaneously. The same 2011 Texas cold snap that tripped off coal plants also prevented dozens of natural gas plants from coming on to back up the coal plants, triggering statewide rolling blackouts.[37] And as the major gas storage leak at Porter Ranch in Aliso Canyon, Calif., recently demonstrated, natural gas has a tendency to float away if you're not careful. That leak prompted the state's grid operators to warn of potential blackouts.

---

37 A deeper freeze in 2021 caused statewide natural gas production to drop 50%, knocking a significant fraction of the state's natural gas power plants offline.

There are countless examples, but the key point is that every fuel and technology has its failure mode. Some fade often but elegantly, like solar and wind whose output gracefully moves up and down through the day and year, while others drop off seldom but suddenly, like coal, nuclear and natural gas.

Rather than obsessing over the variability of renewables, let's acknowledge that all power sources have tradeoffs, with some benefits and risks.

Because of these challenges, grid managers should pursue a suite of options, much the way financial advisers recommend a range of investments rather than putting your life savings in just one stock.

We started with a fable, so perhaps we should draw a lesson in this: After the 2011 blackouts in Texas when coal plants went offline and natural gas plants couldn't pick up the slack, the CEO of the Texas grid operator wrote a public letter to the wind industry, thanking it for feeding enough electricity to the grid to help prevent an even bigger fiasco. The strong oaks of traditional power failed, but the weak reeds of renewables carried on.[38]

---

38 For the February 2021 winter storm in Texas, solar power performed better than expected for extreme weather scenarios; hydroelectric power overperformed; and wind power lagged. Nevertheless, they all bounced back quickly. Additionally, according to official after-action reports by the Public Utility Commission of Texas and the Federal Energy Regulatory Commission, outages by traditional thermal power plants (gas, coal and nuclear) were the biggest problems.

PRAGUE,
CZECH REPUBLIC

Chapter 8

Esteem Needs II: Energy & Innovation

Innovation Should Know No Boundaries

MECHANICAL ENGINEERING, JANUARY 2019

> *American industry – especially the energy industry – is stronger when it can draw talent from all over.*

IRAN, IRAQ, LIBYA, SOMALIA, SUDAN, SYRIA AND YEMEN: THOSE seven countries were singled out when President Donald J. Trump issued a travel ban via executive order in January 2017. The countries on the list were either racked by war or long isolated by sanctions, so it would be easy to believe that the ban would have only minor impact on American industry or the nation as a whole.

And yet, travel barriers don't only affect immigrants, asylum-seekers and refugees, but also world-class scientists looking to improve the human condition. The movement of people is valuable to both individuals and to the global economy, and that's even truer for scientists and researchers, whose endeavors usually depend on – or at the very least benefit from – collaboration.

On American university campuses, the travel ban was immediately felt. At the University of Texas in Austin, where I'm a faculty member, 110 students, scholars and faculty were affected – students I know. This isn't some remote, abstract mystery to me: It's real. The impact of that experience can be multiplied across hundreds of universities and every state.

In fact, scientists who work and live in Europe but who have also conducted research in the targeted regions can no longer easily come to the U.S. to collaborate and share insights. Instead of a quick entry into the country, some British researchers have reported months-long delays in obtaining entry visas.

There are also slowdowns and reductions in issuances of H1B visas for skilled legal immigrants, which is a negative hit for industry. For the tech sector and energy industry, which are composed of many global companies, these restrictions and travel bans are problematic. It makes it harder for them to recruit from around the world or work seamlessly within their multinational enterprises. It's possible that an Iraqi supervisor working in the oil industry in Houston could oversee workers in Europe, but would not be able to visit them in person because of the fear that he or she would not be allowed to return to the U.S.

Some people might shrug this off as problems only for foreigners and a potential boon for native-born Americans, but we all benefit from international collaboration. In 2017, the *Economist* magazine noted that all of the world's restrictions on the free movement of human talent hacks $78 trillion off global GDP.

The story of the lithium-ion battery shows what is at stake if the exchange of scientific ideas is disrupted. The Li-ion battery is ubiquitous, carried in our hands inside our smartphones and laptop computers, and it promises to enable critical opportunities like electric vehicles and the integration of wind and solar into the electric grid. But this breakthrough would not have happened without interactions between different scientists in different countries. The early development of the Li-ion battery spanned from the 1960s through the 1980s and included scientists who interacted and exchanged ideas in Europe, Japan and the United States working at companies like Ford, Asahi, Sony and ExxonMobil, and at universities such as MIT, Oxford and the University of Texas at Austin.[39]

Had policies prevented collaboration between those countries in the 1960s, '70s and '80s, then it is likely that the Li-ion battery would never have come to fruition – at least on its present schedule or with its current excellent performance.

Similarly, new offshore wind turbines and gas turbines are developed by international teams distributed across the globe. Which world-changing technological breakthroughs of tomorrow are we going to miss because of the travel restrictions today?

Because universities are so severely impacted, it makes perfect sense that they were leading the charge to reverse the ban. The lawsuit against the travel ban was led by the state of Washington on behalf of its flagship university.

Universities have been leading voices against the recent wave of travel restrictions, but they should not be alone. It is in the interest of American industries – especially the energy industry – that the most talented workers can come to the U.S. and promote innovation here.[40]

---

39 Nine months after this article was published, this international team whose innovations facilitated the development of the modern lithium-ion battery were awarded a Nobel Prize. One of the laureates, Prof. John B. Goodenough, was my colleague at the University of Texas at Austin and was the oldest recipient of the Nobel Prize in history.

40 On inauguration day on Jan. 20, 2021, President Joe Biden issued a proclamation to revoke the travel bans issued by the Trump Administration. This revocation improved the ability for people to cross the U.S. border, though significant delays for visas remained.

Michael E. Webber

The Case for Cross-Sectoral Disruption

MECHANICAL ENGINEERING | AUGUST/SEPTEMBER 2021

Having a team that can draw on a variety of experiences and points of view can make the difference between being a disruptor and being disrupted.

WHEN THE INNOVATIONS COME FROM UNEXPECTED DIRECTIONS, we need as many points of view as possible. Personal transportation has been a great liberator for modern society. But the resultant downsides – from the national security implications of oil imports to the health impact of air pollution and now the global threat of CO_2 emissions – have inspired all sorts of policy and market solutions that are intended slowly but surely to remake the transportation sector.

Recently, though, transportation has undergone a seismic shift. But it wasn't due to enacting a familiar efficiency measure or traffic planning solution. Individual auto ownership in the United States seems to have peaked due to the widespread adoption of the smartphone.

I'm writing this column from Paris, where hailing a cab is a mysterious art – especially at off hours, in bad weather, or on quiet residential streets. In 2008, two North American travelers in Paris discovered this problem themselves on a snowy night and resolved to fix it. Their timing was fortuitous: A year earlier the iPhone made its debut and a few months before these stranded travelers struggled to hail a cab, the Android smartphone platform was launched. Unbeknown to anyone, the smartphone, with an embedded GPS chip that can conveniently track location, was a transformative transportation technology.

In March 2009, these two weary-travelers-turned-problem-solvers, Garrett Camp and Travis Kalanick, formed a startup that they thought would make it easier to hail a cab. Their company, Uber, did more than that: By providing the convenience of point-to-point trips without the necessity of having one's own car, the company disrupted the century-old concept of individual auto ownership. The company is so successful that its name is now a byword for a whole slew of mobility services.

The expert conversation in the 2000s about how to disrupt the transportation sector completely missed the handheld technology that actually accomplished the feat.

It is not unusual for innovations from one sector to benefit another one. For instance, advanced transportation platforms can help spread innovations in other sectors. Boeing's Dreamliner aircraft helped popularize electrochromic windows, which use electricity to control their opacity. With a touch of a button passengers can change the window's transparency from fully opaque for sleeping and varying degrees of translucence. This feature replaces clunky, pull-down shades, thereby saving weight and giving passengers more control over their brightness at their seat. While futurists had been predicting that similar windows would be adopted in homes and offices because of their obvious advantages, it took a demonstration in the Dreamliner for the technology to seem real enough to spread.

That same plane incorporated carbon fiber, helping foster other industrial applications of the lightweight materials for use in vehicles and elsewhere. In this way, aviation's need for high-performance windows and fuselages helps energy efficiency in our buildings and cars.

There are many examples of this kind of phenomenon. Most famously, space exploration helped create the microelectronics industry, which helped create the modern telecommunications industry, which helped reinvent every other part of society.

The lesson goes far beyond transportation or technology. When we spend too much time on a problem, we often become deeply invested in the pros and cons of a set of solutions that addresses one aspect of the challenge. With our noses pressed so close to the battle lines, we miss the abundance of alternatives that can be clearly visible to others.

That's one reason why diverse teams and broad-based capabilities are not just "nice," but necessary to accelerate innovation. When a sector's biggest disruptions come from unexpected corners, having a team that can draw on a variety of experiences and points of view can make the difference between being a disruptor and being disrupted.

It is also as good a reason as any to broaden one's own horizons and humbly explore other regions and disciplines. The insights we gain might yield surprising outcomes.

Shale Boom Could Fuel Batteries

EARTH MAGAZINE, APRIL 2017

By: Yael R. Glazer, Jamie J. Lee, F. Todd Davidson and Michael E. Webber

> *The status as the most carbon-intensive part of society has put the transportation sector in the target of stakeholders seeking to clean up their act. Electric vehicles remain one of the favored pathways for achieving that goal. The shale revolution could, in fact, prove to be a critical and helpful partner for EVs.*

PETROLEUM-BASED FUELS LIKE GASOLINE AND DIESEL HAVE dominated the transportation sector for the last century – and they still do.

However, policymakers and technocrats are looking toward electric vehicles (EV) to move away from gasoline and diesel. With the support of favorable policies and consumer preferences, EVs, led by Tesla's lineup, are taking off. Already, more than a half million EVs are on the roads in the U.S. and more than a million are on roads worldwide.[41] Dropping internal combustion engines in favor of EVs would seemingly be at odds with the idea of increasing domestic oil and gas production, but the trends come together synergistically in ways that are surprising.

In the last decade, major technological advances in hydraulic fracturing and horizontal drilling have tapped into formerly inaccessible

41 By the end of 2022, there were nearly 17 million electric vehicles on the roads worldwide.

shale deposits around the United States, yielding billions of barrels of oil and trillions of cubic feet of natural gas. Production during this shale revolution has reduced the cost of oil-derived products and natural gas while simultaneously mitigating domestic energy security concerns regarding imported oil and reducing emissions – three major policy priorities set forth over the past few decades.

In that same time, as a result of the shale boom, the U.S. has seen domestic crude oil production increase dramatically. Consequently, U.S. net imports (imports minus exports) were lower in 2015 than they have been since 1970. Additionally, because natural gas has half the carbon content as coal on a per-unit-energy basis, the abundance of affordable gas has displaced the utility of many coal-fired power plants, leading to the lowest emissions from the power sector nationwide in decades. Because of the shift from coal to natural gas for electricity generation, the transportation sector's tailpipes are now a larger cumulative emitter of carbon than the electric grid's smokestacks.

This new status as the most carbon-intensive part of society has put the transportation sector in the target of stakeholders seeking to clean up their act. EVs remain one of the favored pathways for achieving that goal. And the shale revolution could, in fact, prove to be a critical and helpful partner for EVs in two ways: First, EVs have to be charged with electricity produced by some means, and many do so with power from natural gas-fired power plants. Second, it turns out that the wastewater from shale production in certain regions of the U.S. has relatively high concentrations of lithium, which could be recovered and used to make the lithium-ion batteries necessary for EVs.

Is There a Lithium Shortage?

Lithium is a naturally occurring element facing growing demand due to its usage in lithium-ion batteries – the type of batteries generally used in EVs, in portable electronic devices like smartphones and tablets, and for bulk grid storage. In fact, very recently, the price of lithium has soared as a result of growth and the anticipated increased demand of products that depend on lithium-ion batteries. The price per metric ton of lithium carbonate imported to China (where many lithium-ion batteries are made) saw an increase of approximately 50 percent from 2009 to 2015; then it more than doubled within just two months in 2015. If demand continues to increase, it is likely that the price of lithium could follow suit in other areas of the world unless additional supplies are brought to market.[42]

Much was written in 2016 about a looming lithium shortage, but there doesn't seem to be a shortage in lithium supply worldwide. In 2015, lithium demand was approximately 32,500 metric tons[43], a tiny fraction of global lithium reserves – the amount considered currently economically recoverable – which the U.S. Geological Survey (USGS) estimates to be about 14 million metric tons.

According to USGS, Chile holds the world's largest reserves with 7.5 million metric tons, followed by China, Argentina and Australia. Identified lithium resources – a measure that includes both proven reserves and stocks inferred to exist – totals about 40 million metric tons. Bolivia holds the most with almost 23 percent of the world's lithium supply (more than 9 million metric tons); Chile also holds a significant portion. The U.S. holds 38,000 metric tons of lithium reserves and 6.7 million metric tons of lithium resources.

In 2017, there was enough lithium produced to meet demand. But things get sticky when we start looking at future demand or worry about

42 Unsurprisingly, the markets responded to increase the supply of lithium while battery manufacturers improved designs to require less lithium. Stock prices for lithium producers crashed on Nov. 15, 2022, after Goldman Sachs and Credit Suisse noted that they expected lithium supply to far outstrip demand by 2025. Source: https://www.marketindex.com.au/news/lithium-stocks-smashed-after-bearish-notes-from-goldman-sachs-and-credit [Accessed Nov. 16, 2022]

43 Lithium supply in 2022 was 130,000 metric tons Source: https://www.statista.com/statistics/606684/world-production-of-lithium/ [Accessed July 18, 2023]

whether the lithium comes from countries that undermine American foreign policy.[44]

Domestic sources of lithium – including lithium extracted as a byproduct of hydraulic fracturing – might help ensure the secure and steady supply that is needed.

Oil and Gas Industry's Lithium Supply

By geologic fortune, some of the country's major shale oil- and gas-producing resources naturally contain significant amounts of lithium, some of which comes to the surface in produced water along with mined oil and gas.

Produced water generated from shale oil and gas operations contains numerous constituents, including lithium, the concentrations of which all vary by region, depending on geology. Some, like naturally occurring radioactive materials, present clear environmental and health hazards. High salt and total dissolved solids (TDS) concentrations in produced water are also hazardous. TDS levels in produced water can be upward of 250,000 milligrams per liter in places like the Bakken in North Dakota – nearly an order of magnitude more saline than seawater at about 35,000 milligrams per liter. These elevated concentrations present environmental hazards in the event of spills.

Because of the often-hazardous composition of produced water, the predominant management strategy in most regions is to sequester it underground via deep well injection. However, because deep well injection has been linked to earthquakes in areas not traditionally considered seismically active – such as Oklahoma and Ohio – and because other shale-producing regions, like Pennsylvania, have few permitted

44 Much is still being written about supply and demand issues. A 2023 analysis noted supply is insufficient to meet growing demand. Source: https://www.spglobal.com/commodityinsights/en/market-insights/latest-news/metals/040423-global-lithium-demand-seen-outpacing-production-in-2023-oce# [Accessed July 19, 2023.]

injection sites, identifying new management methods for wastewater from shale oil and gas production is critical.

There are other options for the handling of wastewater from oil and gas extraction, some with potential benefits. For example, local communities could use the salty, produced water to de-ice roads. Cleaned-up wastewater could potentially be used for irrigation, or even discharged to surface waters where permitted by regulation. The mineral-rich water could also potentially be a source of lithium.

Is Mining Lithium From Produced Water an Option?

Just how much lithium is contained in produced water is hard to know, given that oil and gas operators are rarely required to report water-quality data about their wastewater. USGS does make some information public, however. From this limited dataset, along with data from our own research with various oil and gas partners, we estimate lithium concentrations in produced water from across the Marcellus Shale region – stretching across much of Appalachia and into the Northeast – of approximately 80 to 200 milligrams per liter.

In 2014, the volume of produced water in Pennsylvania (presumably mostly from the Marcellus Shale but also from the Utica Shale) alone was approximately 29 million barrels, or 1.2 billion gallons. If we consider 120 milligrams of lithium per liter as a typical concentration for the region, we calculate that Pennsylvania's wastewater contained about 545 metric tons of lithium.

It's highly unlikely that all the lithium in produced water could be economically recovered. But even with a 60 percent recovery rate – an estimate based on published research detailing methods and recovery rates for lithium extraction from brine waters – the produced water from Pennsylvania alone could potentially supply approximately 8 percent of Tesla's annual lithium needs at its Gigafactory in Nevada once it reaches full-scale production. That's not a trivial amount. This percentage could increase if produced water in other oil- and gas-producing shale regions in the U.S. also contains lithium.

The remaining question is how feasible – both technically and economically – it is to extract lithium from the produced water of oil and gas operations. Based on available published research, we are confident that it is technically feasible. However, it is unknown how expensive it would be to extract lithium from produced water from oil and gas wells.[45]

Currently, lithium is often extracted from underground brine pools, which have immense volumes, unlike the relatively small water volumes of oil and gas wells. The economies of scale for traditional lithium harvesting bring the extraction cost down and makes the process economically viable. Due to the distributed nature of supplies of produced water – coming from numerous oil and gas wells not always owned by the same operator and spread over large areas – it remains to be seen if using existing extraction methods is economically feasible at current lithium prices. As oil and gas producers move toward wastewater treatment and reuse to avoid expensive disposal costs, it will be interesting to see if it becomes possible to extract the lithium for minimal additional cost.

If lithium prices continue to increase, though, it might not be long before separating lithium from produced waters becomes attractive even without other developments. If that happens, shale production would not only be a source of natural gas to produce electricity with which to charge batteries, but could become a domestic source of critical minerals for manufacturing batteries for electric vehicles.

45 By 2022, there were several startup companies and industrial efforts dedicated to extracting lithium from brines.

Michael E. Webber

Breaking the Energy Barrier: Can the Department of Defense Solve the World's Energy Crisis One Jet at a Time?

EARTH MAGAZINE, SEPTEMBER 2009

The Department of Defense, the world's single largest energy consumer, is much more involved with energy issues than people realize. Because of its great need for energy, the DOD is also increasingly becoming a leader in energy research and development, and in implementing new energy technologies and using alternative fuels.

THE UNITED STATES HAS ALL THE TOOLS IT NEEDS TO SOLVE THE country's energy problems – and one of the surprising contributors to the solution is none other than the U.S. Department of Defense. Although many look to the Department of Energy to take the lead on energy issues, one of the ironies in the federal government is that, contrary to its name, the Department of Energy historically has been a part of the national security apparatus; it's a weapons agency that happens to dabble in energy on the side.

The unexpected corollary is that the Department of Defense is much more involved with energy issues than people realize. The department is the world's single largest energy consumer, a guarantor of energy trade, a victim (and beneficiary) of energy-related military tactics and a strategic protector of U.S. interests in energy-rich areas of the world. Because of its great need for energy, the Department of Defense is also increasingly becoming a leader in energy research and development, and in implementing new energy technologies and using alternative fuels. Therefore, the department is a top contender for finding a low-carbon, domestic source of energy that is compatible with existing infrastructure and that satisfies the country's energy needs just as well as traditional forms of energy.

If the Department of Defense can do that, then it will not only solve the United States' energy woes, it will also go a long way toward solving the world's energy challenges.

A Major Energy Consumer

The Department of Defense is responsible for more than 80 percent of the U.S. government's energy use and approximately 1 percent of the nation's annual energy use – more than any other single entity. In 2008, the department consumed approximately 900 trillion Btu of energy[46] – more than the entire annual consumption (at the time) of countries such as Bulgaria, Denmark, Israel and New Zealand. The Department of Defense uses its fair share of electricity and requires natural gas, coal and steam for heating, but most of its energy appetite is for liquid fuels: 120 million barrels (5 billion gallons) per year. Jet fuels alone account for more than two-thirds of that liquid energy – fueling the U.S. military's unparalleled mobility and allowing the country's armed forces to establish their presence anywhere in the world. But being a major energy consumer is a double-edged sword: On one hand, it gives the Pentagon a lot of experience with the capabilities and limitations of different energy sources, but it also means the Pentagon is responsible for most of the more than $12 billion the U.S. government spent on energy in 2007.

In addition to the price tag for fuels, the Pentagon spends a significant amount of money moving that fuel around. For example, refueling the Air Force's planes in midair increases fuel costs to about $42 per gallon. Delivering fuels on the ground to the battle lines costs hundreds of dollars per gallon. The Army alone devotes 60,000 uniformed personnel to transporting fuels worldwide. On top of that, the number of fuels is proliferating along with new technology, doubling from about a

46 Department of Defense energy consumption dropped to 650 trillion Btu by 2021, primarily by significantly reducing jet fuel consumption and fuel oil consumption from reduced military activities overseas.

half-dozen different battlefield fuels two decades ago to more than a dozen today. This means the military has to pay even more attention to the logistical details of moving fuel. Because energy use has a critical impact on the Department of Defense's bottom line, the department has a much greater incentive to find ways to reduce its consumption than the Department of Energy or any other government agency. And because agency budgets are often a zero-sum game, if the Department of Defense can reduce its bill for fuels, then it might have more money available for other purposes. Thus, the military, more than any other segment of government, has a direct self-interest in finding an affordable and reliable supply of energy.

An Energy Protector

One looming energy challenge is that the world's demand for liquid fuels is projected to grow significantly in the coming decades (from approximately 84 million barrels per day today to more than 110 million barrels per day in 2030[47]), yet the available supplies of conventional energy sources are becoming increasingly constrained. About two-thirds of the world's remaining conventional petroleum and natural gas resources are in the countries of the Middle East and the former Soviet Union. Because energy is such a critical economic and political concern for the United States and its allies, keeping the hydrocarbons flowing is a top strategic priority.

As a result, the U.S. military expends a significant portion of its energy stabilizing or protecting energy-rich parts of the world. The military also gives a show of force to keep the world's avenues of energy commerce open and freely flowing to ensure a stable global economy. Consequently, the U.S. Navy regularly patrols critical energy chokepoints such as the Strait of Hormuz near Iran. And AFRICOM, the U.S. military force created by the Pentagon in 2007 to manage operations in

47 On track with that projection, global daily consumption of liquid fuels hovered right at 100 million barrels per day in 2022.

Africa, spends a great deal of time stationed off the coast of West Africa, the continent's main oil-producing region.

It's hard to put a price on this total effort, but different analyses estimate that the United States spends somewhere between $29 billion and $143 billion every year just to protect the supply and transit of oil. These expenses mean Americans in 2009 are paying anywhere from about $9 to $44 of additional hidden military costs for each barrel of petroleum we import from countries other than Canada and Mexico.

All of this energy-inspired intervention worldwide gives many planners in the Pentagon a grave philosophical headache when they realize that we spend significant amounts of energy to stabilize the very regions from which we import those energy resources in the first place. Thus, many in the Pentagon are beginning to ask whether there's a better way to meet our energy needs that doesn't require such an exacting toll on money, personnel and energy.

The Leverage

Thankfully, the Department of Defense has some critical policy levers at its disposal to solve its own energy problem. It has firsthand experience producing, distributing and using energy, which comes in handy when developing alternative fuels. In addition, the Department of Defense has a unique ability to enter long-term purchasing agreements for power (20 years or more) and fuels (up to five years), which gives energy developers the revenue stream they need to secure financing to bring large-scale projects online. In today's tough economic climate, where the inability to obtain financing hobbles many good energy projects, this purchasing power is a valuable asset and can be used to get new biorefineries or novel fuel-production systems over the hump from blueprint to reality.

Perhaps the most important tools at the Department of Defense's disposal are its significant research budgets, including those for the Defense Advanced Research Projects Agency (DARPA) and the research arms of the different military branches such as the Office of Naval Research and the Air Force Office of Scientific Research. The total amount of R&D money related to energy in the federal government's research portfolio has lagged behind health- and defense-related research. Based on the sheer size of investments, the Department of Defense has more capacity to conduct effective energy research than all other government agencies combined.

After decades of ceding that bureaucratic turf to the Department of Energy, the Department of Defense is starting to invest its own money to tackle the energy problem.

Higher Standards

The Department of Defense clearly has the incentive and wherewithal to lead the transition to alternative fuels, but there's another reason why the department is in a prime position to move society forward on the energy front: It has stricter performance requirements and thus its energy challenges are harder to solve.

When it comes to fueling planes and tanks, the U.S. military needs higher-performing fuels than the Average Joe filling up his truck at the corner gas station. The Air Force in particular has special energy needs. Because weight is at a premium when flying, airplane fuels must be very energy dense. These fuels must also have convenient boiling points to ensure they work in the scorching heat on a tarmac in Iraq and convenient freezing points so they can continue to work in high altitudes. These requirements make ethanol and electricity stored in batteries impractical for many aviation applications because they both have much lower energy densities than petroleum-based jet fuels. Consequently, the Department of Defense will have difficulty using conventional, off-the-shelf alternative energy solutions from the ground-based transportation sector to meet its needs.

To complicate matters even more, the Air Force has set an ambitious goal for itself: to fuel half of its North American fleet with domestic fuels made from alternative feedstocks by 2016.[48]

48 I have not found evidence that these goals were met.

One choice is to use the country's vast coal resources to make synthetic fuels via the Fischer-Tropsch process, which chemically converts hydrocarbons into custom synthetic blends that can replace gasoline, diesel or jet fuel (also known as coal-to-liquids, or CTL). The Air Force intends to certify its entire fleet on 50/50 blends of petroleum-based jet fuel and CTL synfuels by 2011 and is ahead of schedule:[49] Several planes, including the C-17, B-1 and B-52, are all certified, and the Air Force is already testing the KC-135, F-15, F-22 and the midair refueling fleet. To conduct these tests, the Air Force already consumes more than 300,000 gallons of CTL-based jet fuels annually.

These tests show that CTL fuels work well, but the Air Force experienced a bump along the road to expanding its use of CTLs: the Energy Independence and Security Act of 2007. The act includes a provision targeted at the Air Force, which prohibits it from using CTL fuels for anything other than R&D purposes unless the fuel's life cycle greenhouse gas emissions are lower than conventional petroleum. Right now, the life cycle greenhouse gas emissions of CTL fuels are higher than jet fuels made from conventional petroleum.[50]

Although the act may potentially hobble the development of domestic CTL fuels as a replacement for petroleum, it also opens the door for new biofuels. Ethanol does not have a desirable energy density for aviation applications, but jet fuels similar to biodiesel can be made from a few different processes using a range of feedstocks such as soy, palm oil or algae.

Researchers have known for decades that algae are a source of fuel, but developing algae-based fuels has been plagued by prohibitive cost and scale hurdles. That's where DARPA – the R&D arm of the Pentagon

49 In 2021, the U.S. Air Force partnered with a company to make e-fuels (liquid fuels made from electricity and carbon dioxide from the atmosphere that can be used as a direct substitute for petroleum fuels).

50 Because of the greenhouse gas emissions impacts from coal-to-liquids (CTL), the Air Forces scuttled its plans for a CTL plant.

famous for creating the Internet – comes in. DARPA invested upward of $50 million for a crash-course, 24-month program to generate thousands of gallons of algae-based jet fuels at a price competitive with gasoline. It just might be that DARPA will move algae across the threshold of scalability and affordability.[51]

Into the Wild Green Yonder

In the end, the Department of Defense has a large responsibility to help the world avoid energy crises – and it does so by stabilizing energy-rich regions and preventing supply cutoffs. But it comes at a huge cost in terms of energy, money and personnel. This complex relationship with energy, combined with the department's purchasing power, gives it unique motivation, insight and capability to solve the world energy problem. And, in fact, it's already doing more than most people realize. The United States has spent decades building the most capable military in the history of the world, and energy just might be one more area where it's poised to excel.

SIDEBAR

The Department of Defense Is a Green Power

Beyond transportation fuels, the Department of Defense is also taking the lead on green electricity. Since World War II, the U.S. Army Corps of Engineers has operated many of the United States' largest hydroelectric dams, sources of emissions-free electricity. Today, the U.S. Army Corps operates 75 hydroelectric power plants across the country that together generate more than 2 percent of the nation's electricity.

The U.S. Navy, meanwhile, has relied on small-scale, distributed nuclear generation as a reliable source of low-carbon power for decades. This experience might open the door for miniature on-site power plants that would allow domestic military installations to go off the grid and

51 To date, algal biofuels still are not commercially available.

increase resilience against vulnerabilities to the grid. Some military installations, such as Forts Bliss, Hood and Carson, have announced various net-zero initiatives.

In addition, the Department of Defense has embraced renewable energy. In 2005, the U.S. Air Force was the largest purchaser of renewable energy in the EPA's voluntary Green Power Partnership, which helps organizations procure green power. The nation's largest solar photovoltaic array in 2009 was a 15-megawatt capacity at Nellis Air Force Base in Las Vegas, Nevada. And a few bases, including Dyess in Texas and Fairchild in Washington, receive 100 percent of their energy from renewable resources.

Research Is Necessary to Accelerate Our Transition to a Zero-Carbon World

POUR LA SCIENCE (THE FRENCH EDITION OF SCIENTIFIC AMERICAN), APRIL 2020

> *Our relationship with energy is about harnessing the benefits and containing the environmental impacts of those transformations. Bringing the benefits of energy to everyone without heating the atmosphere, acidifying the oceans, or denuding land will require new thinking and new solutions.*

BY CLASSICAL DEFINITION FROM THE EARLY INDUSTRIAL ERA, energy is the capacity to do work; however, in the modern context that seems very limited compared with what energy actually offers society. Taking a broader and updated view, energy is the ability to do interesting and useful things. Energy brings illumination, information, heat, clean water, abundant food, motion, comfort and much more to our homes and factories with the turn of a valve or the touch of a button. It is the potential to harvest a crop, refrigerate it, and fly it around the world, but it also guarantees education, health and security. Our civilization is founded on access to energy and the corollary is therefore that a lack of energy would lead to its collapse.

An absence of energy does not mean it disappears entirely: The laws of thermodynamics tell us that energy is conserved. Energy is an inherently finite resource. We cannot make more of it; we can only move it around or transform it. At its heart, our relationship with energy is about harnessing the benefits and containing the environmental impacts of those transformations. Bringing the benefits of energy to everyone without heating the atmosphere, acidifying the oceans, or denuding land will require new thinking and new solutions.

The World Is a Complex Place

In the modern world, everyone uses energy with a mixture of elation and guilt: It's the energy-lover's dilemma. How do we get all the benefits we enjoy from consuming energy without the downside impacts of pollution, price volatility and national security risks?

The answer is to recognize that each fuel and technology has its upside benefits and downside risks. Worldwide energy use is a complex system with many parts. The energy sector is intertwined with society in many obvious and nonobvious ways.

If there is one lesson we should keep in mind for our energy challenges, it is that there are no universal, immediate solutions. We need a suite of solutions adapted to each location because no single option can get us all the way to our zero-carbon future without a critical drawback such a high cost, insufficient scale or poor reliability.

That means we need innovative thinkers to develop more options, lower the cost of existing options and to optimize how they all fit together. The world's research ecosystem of industrial facilities, national labs and universities in many countries must step up with high levels of internal and governmental support to accelerate the pace of our innovation. And because the challenge is complex and too big for any one company or government to solve, we must collaborate across sectors, academic disciplines and borders.

The way to solve this conundrum is not by hashing out old clichés of fossil fuels versus renewables, electricity versus gas, or other tired battles. We need a more refined view. The same thinking that got us into these problems – drill more, pave more, consume more – will not get us out of them. Tired techno-enthusiasm that just says we can use smarter gadgets won't be enough, either. Energy efficiency is one way of reducing our carbon footprint without affecting our lifestyle, but it is not enough.

The fastest, cheapest and most reliable way to reach zero-carbon energy includes a mixture of low-carbon electricity and low-carbon gases. We need cleaner forms of energy and we need to clean up conventional forms with carbon capture and scrubbers so that we can maintain and expand energy access without scorching the planet.

The areas that need rapid attention are low-carbon power generation (such as wind, solar and geothermal energy), low-carbon gases (such as biomethane, synthetic methane, hydrogen and hydrogen carriers such as ammonia, formic acid and methanol), technologies that reverse the accumulation of CO_2 in the atmosphere (through carbon capture, direct air capture and soil carbon sequestration), cross-cutting tools (such as drones, robots, sensors and artificial intelligence), and smart and efficient uses of energy (including energy storage, smart appliances and user education to change our behaviors and habits).

ENGIE's[52] corporate research program is organized around these themes and our benchmarking with the world's largest national laboratories and energy and technology companies indicates that we are not alone with our view of the future. Indeed, most of us are tackling similar problems. We just need to move more quickly collectively.

Adopting a cleaner suite of options while increasing energy access and letting go of our dirtier past is our critical path forward. Change is

52 ENGIE, headquartered in Paris, France, is one of the world's largest energy and infrastructure services companies. From 2018-2021, including when this article was authored, I served as ENGIE's Chief Science and Technology Officer.

good, so we should go for it. But change is slow, so we better get started. This is where we need research: to accelerate the transition.

Time Is Running Out

It usually takes decades or centuries to transition from one dominant fuel or technology to another. In the United States, coal became the most popular energy source in 1885 and was only surpassed 65 years later by petroleum in 1950. Petroleum still leads today, but might be overtaken by natural gas in the next decade, meaning it will have ruled for 80 years.

While natural gas provides some important environmental and performance benefits, we don't have another 80 years to wait to replace it with cleaner options such as zero-carbon electricity or other gases with lower-carbon footprints. That means the race is on and our task in the research world is to increase the scale and decrease the cost of those alternatives so they can be adopted widely sooner.

The mission of our teams is now clear, we just need to pick up the pace to meet the energy challenges alongside out scientific and academic partners and develop the energy solutions of the future, the ones that will allow us to preserve biodiversity, the climate and social inclusion.

Build First, Explain Second

MECHANICAL ENGINEERING, JUNE/JULY 2021

> *We can ramp up on deployment of clean-energy systems, and then apply what we learn to make the whole system better.*

WHEN MARTIN EBERHARD AND MARC TARPENNING – MEN WITH backgrounds in electronics and software – announced in 2003 that they were starting a company to make electric vehicles, it was easy to dismiss them. After all, neither was an automobile researcher or had worked for carmakers. Instead, it seemed that Eberhard and Tarpenning wanted to make cars first, and then learn how to make them later.

In the modern global economy, we have this organized and elegant vision of how innovation and commercialization unfolds: It goes from science to engineering to manufacturing to market. It starts with scientists who propose new theories that are then experimentally verified. These tests lead to ideas for applied scientists to demonstrate, after which engineers improve the concepts until a company manufactures the widget for the benefit of humanity.

For the present energy transition away from carbon fuels, this whole process from fundamental science to lab to pilot to scale is too slow because it can easily require several decades. It needs to start immediately.

Fortunately, there is another way.

In fact, in the Industrial Revolution, innovators went the other way around: They built steam engines and then afterward the underlying scientific principles of thermodynamics were developed to explain how the machines worked.

Inventors such as Savery, Newcomen or Watt knew that higher pressures and temperatures could achieve higher power output and that tighter boiler designs would reach higher efficiency, but they did not know why that was the case. As historian Bruce Hunt wrote in his book, *Pursuing Power and Light*, "Historians of science and technology have often and quite rightly observed that the steam engine did far more for science than science ever did for the steam engine."

Despite the initial lack of accurate theories and scientific applications, these machines transformed the world and facilitated a lot of follow-on innovation in a short period of time.

Perhaps it's time for us to go backward again. Instead of developing new theories about how to decarbonize society in the most elegant forward-looking way, let's get in the field and just start building low-carbon solutions and then go backward to figure out which ones worked and why.

It takes too long to follow the normal process, so we need to switch the order to accelerate things. Build first, explain second.

The multitrillion-dollar infrastructure bill is an opportunity to put this idea to practice. Many people use the lack of perfect science on climate change, ocean acidification, geoengineering and other fields as an excuse for delay anyway, so we might as well get going.

Looking back over recent history there's reason to be confident this approach would work. After all, the costs for wind, solar and batteries did not drop swiftly over the past decade or two because our theories improved; their costs dropped because we installed a lot of them and the supply chains improved and economics of scale were reached.

Research is still important – we need it to identify new battery chemistries that are less prone to thermal runaway and composed of solid-state materials that are stable and abundant; to reduce degradation of solar panels; to finetune wind turbine performance through a range of conditions; and to develop carbon capture or nuclear designs that overcome their cost burdens and safety risks. This research should continue, but let's not use our slow learning process to delay urgent action.

We could emulate Eberhard and Tarpenning, who together with Elon Musk turned Tesla into the company that is setting the future direction for the automotive industry. Let's ramp up on deployment as fast as we can, and then apply what we learn to make the whole system better. We just might surprise ourselves with how quickly we can install solutions. Along the way we'll employ a lot of people, overcoming fears that decarbonization will be bad for the economy. It's possible that going backward will be the fastest way forward.

Energy Is All Around Us, Including Up Above and Down Below

POUR LA SCIENCE (THE FRENCH EDITION OF SCIENTIFIC AMERICAN), MAY 2021

> *Just as energy opens up the doors to new frontiers, those new frontiers give us energy.*

ONE OF THE ENDURING LEGACIES OF ENERGY IS THAT IT IS ALL around us, hidden right in front of our eyes. People know that energy brings light to darkness and warmth to the cold. But despite its ubiquity, it remains hidden. Ask most consumers where energy comes from and they will point to the light switch or outlet on the wall, or maybe to the gas station down the street where they purchase fuel for their car. Indeed, the entire system that brings energy from its raw form somewhere far away to a useful form such as gas or electricity in our homes is a mystery. The fact that energy shows up quietly and regularly just when we need it gives it an almost magical quality. This magical aura reminds me of the source of my original desire to become an engineer: space travel.

Drawn to the Stars

The magic of space travel inspires joy in many people, so I was not alone. For my undergraduate degree, I decided to study aerospace engineering. I had ambition to be part of the space program, not as an astronaut—that heroic job belongs to people who are not afraid of heights and enjoy roller coasters—but as one of the engineers who would help take humanity to new frontiers.

I was fortunate to spend two summer internships at NASA's Ames Research Center in California where I worked on supersonic propulsion. I continued my interests in space exploration while developing sensors as part of my doctoral research. The first sensor I designed was for detecting fuel leaks on the launchpad of Kennedy Space Center in Florida. The second sensor was used to monitor a water recycling system onboard the International Space Station, which we tested at the Johnson Space Center in Texas.

While these experiences were related to the space program, energy was at their core. Modern fuels were the critical ingredient for space propulsion. The safety risks of leaking fuel on the launchpad motivated my work to develop a sensor. And the incredible energy burden of lifting freshwater to space motivated the desire for an onboard water recycling system.

I finally realized that modern forms of energy are central to our pursuit of the heavens.

That's when I switched career directions to focus on energy. Energy gives us the ability to push our boundaries further.

The Magic of Energy

For older forms of transportation, it was wind with sailboats or muscle power rowing large Viking ships that freed humans to explore the world. As fuels improved, coal to power steamships or trains, diesel and gasoline for cars and trucks, jet fuel for planes, and ultimately rocket fuel for space travel, we could travel farther and faster. Energy is the magic key that unlocks the doors to these distant locations and, as time goes by, we get there more quickly and more often.

But our relationship with energy is more complicated than that. Just as energy opens up the doors to new frontiers, those new frontiers give us energy. The most difficult frontiers today remain space, the deep ocean, and below Earth's crust. In a symbiotic partnership, we use energy to explore beneath the land and the ocean's surface and then bring energy back up. The fuels that took us up to space came from down below. The pollution we put up into the atmosphere can be sequestered down below. And space is the test bed for our latest technologies such as fuel cells and thermoelectric generators. The future will connect these disparate systems more closely.

Going to space also unlocked a new vision: For the first time, astronauts could look back down and see the Earth in its entirety. It is no accident that the Apollo program in the late 1960s coincided with the peace and environmental movements in the United States. From space, the absence of borders between countries is obvious, which makes war seem unnecessary. And the beauty of the planet helped foster more attention to protecting its fragile ecosystems. Energy enabled this global view.

There are two inescapable facts from space. First, most of the planet is below the land and oceans. To learn more about our home, we must go deeper. Second, the entirety of Earth shares a single atmosphere. This fact was already known, of course, but from space the obviousness of the shared skies is hard to avoid. What we have come to realize very sharply in the last few decades is that energy's pollution spreads globally through this common atmosphere.

Changing Perspective

Environmental concerns from prior eras were local in nature. Water contamination would happen from a nearby mine. Air pollution would cause asthma in the factory town or acid rain in a neighboring region. But because greenhouse gases like carbon dioxide are long-lived, stable and mix rather uniformly in the atmosphere, climate change today is happening on a worldwide basis. How do we increase access to energy for those 1 billion people who will suffer from climate change, but who do not have modern lifelines such as electricity, piped water or sanitation?

Can we increase their access while decreasing the global climate effect of the other 7+ billion who already have access? How do we change an industry active in every country and whose no longer isolated impacts are endured worldwide?

The problems are not easy, so we will have to look for answers in new places. Going deeper or going higher will require more innovation, but it will also unlock new potential. New frontiers demand more technical excellence from us, but also hold some of the solutions we need.

Section V

Self-Actualization Needs

CASTLE COMBE, UK

Chapter 9

Self-Actualization Needs I: Energy & Society

Redefining Humanity Through Energy Use

EARTH MAGAZINE, MARCH 2010

> *I contend that what really separates humans from all the other species is that we are the only ones to manipulate energy.*

WHAT IS IT, EXACTLY, THAT DISTINGUISHES US FROM OTHER species? The definition of humankind has perplexed scientists, philosophers and theorists for centuries. DNA composition differentiates species in a technical sense, but that definition is hardly satisfying. Certainly there must be something more ethereal that separates us from "lower" forms of creatures. Over the centuries, several definitions have emerged – from using tools to speaking – but have then been proven insufficient in some heuristic way. So I propose another option: manipulating energy.

One of the earliest definitions of humankind is that we make tools, while other species do not. The unspoken corollary is that our tools are a reflection of our humanity: The more advanced our tools, the more advanced our civilization. In our modern thinking, the finely honed metal tools of the Middle Ages reveal a more advanced (and civilized) society compared to the rugged creators of ice-age stone adzes. And, subsequently, today's humans, with cyber-tools and sophisticated nanoscale electronic devices, are more advanced than the people from the Middle Ages. But Jane Goodall turned that definition on its head. Goodall discovered that chimpanzees make tools: They fashion leaves into cups, collecting water from pools in the knots of trees, or strip sticks of leaves and thorns, leaving a smooth stem to extract termites from mounds.

Another proposed distinction is that humans have language, whereas other animals do not. Language was the gulf that separated us from other animals – until Koko the gorilla came along. Born in 1971, Koko was taught approximately 1,000 signs of American Sign Language. Although proficient signing is not quite the same as communicating via language, there were some striking incidences where she made her point clear. She created insults, including classics such as "stupid devil," "devil head" and "you dirty bad toilet." She also invented words when she lacked the proper vocabulary: "finger bracelet" for "ring" and "water bird" for "duck" are just two of many examples. A linguist is unlikely to declare that Koko's imaginative use of sign language qualifies as a language in the strictest sense; however, a casual observer might conclude that Koko's ability to invent words and phrases to convey meaning indicates that the language gulf separating humans from other animals is less robust than we once thought.

In recent years, some people have said the definition of humankind could revolve around sex: Humans control their reproduction and engage in recreational sex; other animals do not. Upon inspection, this definition also fails to withstand scrutiny. Dolphins, mountain goats, rhesus macaques and proboscis monkeys (to name just a few) all engage in various forms of nonprocreative or recreational sex. The full range of activities would make most readers blush.

And so back to the drawing board.

I contend that what really separates humans from all the other species is that we are the only ones to manipulate energy. The First Law of Thermodynamics tells us that energy has many forms (for example, chemical, thermal, kinetic, electrical, atomic, radiant) and that we can convert one form to another. And though all species benefit from the natural conversion of radiant energy (for example, sunlight) into chemical energy (derived from, for example, photosynthesis), humans are the only species that will specifically manipulate energy from one form to another – for example converting chemical energy (fuels) to thermal energy (heat) or mechanical energy (motion).

And, thus, a new definition of humanity is born: Humans intentionally manipulate energy.

The good news is that we can be proud of our ability to use these manipulations as a way to advance the quality and length of our lives. It's human nature to manipulate resources around us to our benefit. The bad news is that because energy consumption inherently has environmental impacts, the corollary is that only humans intentionally pollute. And that is the paradox that remains as the ultimate challenge for humanity: How do we get the benefits of energy without its problems?

Perhaps we can use this new vision of ourselves as a call to action. By admitting that energy is a central and defining aspect of who we are, perhaps we can finally find the courage to accept responsibility for its negative effects. After all, that would be the humane thing to do.

The Color of Energy

MECHANICAL ENGINEERING, MAY 2020

The energy industry must work to solve its racial disparities.

IN MANY RECENT DISCUSSIONS ABOUT THE FUTURE OF ENERGY, the language has been colorful. My environmentalist friends push for green policies and programs with names like "green choice." My friends in industry sometimes push back, noting that the grid is colorblind and cannot distinguish the green electrons of a wind farm from the brown electrons of an outdated coal plant.

Planners differentiate brownfield sites, where existing energy infrastructure is located, from greenfield sites representing new construction on nonindustrial lots. And those of us innovating to facilitate the clean-energy transition will often classify hydrogen as being either green, blue or gray, depending the relative cleanliness of its production and use.

Despite the diversity of color in our language for energy, however, there is a remarkable lack of diversity in the workforce.

After years of highly public diversity campaigns by major companies, in 2019 less than 7 percent of the American workforce in oil and gas extraction identified as Black. Coal mining is even more homogeneous; according to the U.S. Bureau of Labor Statistics, only 4.6 percent of the coal mining workforce identifies as Black, Asian, Hispanic or Latino.

Even starker than the lack of diversity among rank-and-file is the reality of the glass ceiling for energy executives of color. Stats are hard to come by for the energy industry, but a McKinsey & Co. study with LeanIn.org, found that men and women of color make up just 14 percent of C-suite positions across all sectors. Just one in 25 C-suite positions is a woman of color. Central Texas hero Paula Gold-Williams, who leads

San Antonio's electric and gas utility, stands out as the only Black female CEO of a major energy company in the United States.[53]

Why does this matter? For one, people of color globally are the ones who have the least access to energy – the billion people without access to electricity live primarily in sub-Saharan Africa, Latin America and Southeast Asia.

It matters because people of color endure a higher risk from the energy sector's pollution. One study in California found that Black and Hispanic residents suffer risk exposure to environmental hazards nearly six times higher than white residents.

An energy sector that drew more workers and executives from those communities might push to end that unfair status quo.

It's also an economic challenge. As a legacy of decades of racial disparities in transportation and housing policies, the energy cost burden for minority neighborhoods exceeds that of white neighborhoods for a range of wealth levels. There are several reasons. Poorer families more often live with multiple generations in older homes that are leakier (hot air leaks out in the winter and leaks in during the summer) and they are less able to replace inefficient appliances. This combination is a perfect recipe for higher utility bills.

Sometimes poor customers even pay to subsidize wealthy customers. Many municipal utilities incentivize solar panel installation to improve grid reliability and reduce local air pollution. It's usually rich customers who take advantage of the subsidies because they are the ones who can afford solar arrays that easily cost $10,000 to $20,000 per rooftop, even after rebates.

As those subsidies are usually spread across the entire service area, customers who can't afford to install double-pane windows or efficient

53 Ms. Gold-Williams was removed from the CEO position in late 2021 in the wake of statewide energy outages during the winter storm earlier that year. After she was fired, there were zero black women in CEO positions at major energy companies nationwide in 2021 or 2022.

air conditioners are helping to pay for solar arrays on the roofs of wealthy homeowners across town.

All told, it's time for society to reckon with racial and social disparities that live on to this day. And because of the critical role energy access plays for fostering economic potential – and the acute damage caused by energy pollution – the energy industry should play a leadership role in solving this blight. Diversifying the workforce and the leadership teams is the first obvious, visible step to broaden the range of voices in decision-making for critical projects.

Setting the Direction

MECHANICAL ENGINEERING, SEPTEMBER 2019

For women to take a more equal leadership role in the energy industry, policies need to support families.

AT THE GRASSROOTS, THE GLOBAL ENERGY SYSTEM IMPACTS – AND is impacted by – women. In the developing world, primitive cookstoves and fuels such as wood or cow dung produce poor indoor air quality that leads to the premature deaths of millions of women annually. Conversely, when rural areas gain access to electricity for water pumping, it's primarily women who are relieved from the physical burdens of fetching and carrying water. In the developed world, women are often the primary decision-makers when families buy energy-consuming appliances such as refrigerators and air conditioners.

In spite of this, women are largely excluded from setting the future direction of the energy industry. Women make up less than 25 percent of the total employees in the energy industry, and in the oil and gas sector in particular, only 1 percent of CEOs are female.

My perspective as an American expatriate working in France has given me some new confidence that greater gender balance is possible. Women make up the majority of the workforce in France, and at ENGIE, the Paris-based energy and infrastructure services firm where I serve as the Chief Science and Technology Officer, four top executives are women, including the CEO, Isabelle Kocher.[54] This greater degree of parity extends to every part of the organization: At the main corporate research center, for instance, women make up 37 percent of the lab's researchers.

How did ENGIE – and France as a whole – get so far ahead with gender balance among its workers and corporate leaders? Because of other seemingly unrelated but important policies that are family-friendly: maternity leave, parental leave, generous vacation time and accessible – and affordable – daycare and health care.

In France, mothers get 16 weeks of maternity leave, including six weeks before the birth, at full pay. Even more weeks are provided for multiple children. For instance, mothers can claim 26 weeks of paid leave for their third child or up to 46 weeks if they give birth to triplets. (By contrast, U.S. law guarantees only unpaid leave for new mothers.) Mothers – and fathers, too, by the way – also get three years of optional parental leave, which allows them to stay home while they collect a small stipend. While that's not full pay, it's helpful nevertheless.

What's more, the five weeks of paid vacation the French famously revere comes in handy as the kids grow older because it's easier for parents to be available when the children are on holiday breaks from school. The French 35-hour workweek also plays a role, since it means that parents can more readily drop kids off and pick them up from childcare or school. The wider availability (and affordability) of daycare and healthcare also help reduce the stress of balancing work and family life.

Critically, French parents can access these benefits without stigma or repercussions. While mothers are away on maternity or parental leave, their jobs are still protected, so their career is waiting for them when they return to the workforce. Such protection may not entirely eliminate

---

54 Isabelle Kocher was fired as CEO in February 2020, replaced with Claire Waysand as Interim CEO, who was followed in late 2020 by Catherine MacGregor as CEO. As of December 2023, MacGregor was still CEO.

the so-called mommy track for female employees, but it does support women as they progress through their careers.

Thankfully, trends in the U.S. energy industry are turning toward the better. Some female executives, such as Vicki Hollub at Occidental Petroleum and Lynn Good at Duke Energy, are leading major energy corporations. Women also make up 35 percent of the renewable energy workforce globally, so as the industry as a whole continues its move toward wind and solar, it's reasonable to think the gender balance of its workforce might also improve. But overall, progress is slow.

The French experience demonstrates that gender balance in the workforce can be achieved – and importantly, that shattering the glass ceiling in the energy industry is a reachable goal. They also demonstrate that just saying you want women in the workforce isn't enough: Those goals have to be backed up with a whole suite of policies. And it turns out that if you implement policies that are family-friendly in general, they will also be female-friendly and mother-friendly.

Decisions made by the leaders of the energy industry are felt keenly by women around the world. It's time that women are better represented among those decision-makers.

The Farm Woman's Dream, 100 Years Later

THE AMERICAN SOCIETY OF MECHANICAL ENGINEERS ENERGY BLOG, JUNE 29, 2020

> *The Farm Woman's Dream poster is targeted at women, not men, a telling example of who endured the greatest burden from not having access to modernized systems, and who would reap the greatest benefit from modern improvements.*

IN JUNE 1920, THE UNIVERSITY OF MISSOURI ISSUED A POSTER AS part of a campaign to encourage farm owners to modernize their water systems. It is aptly titled "The Farm Woman's Dream" and shows a woman – presumably poor, dressed in ragged clothes – carrying water without gloves from a hand-pumped well in the freezing cold along an icy path uphill toward her house and barn. The poster's text makes the following encouragement:

Make Your Dream Come True
Consult Your County Agent or Write to Your College of Agriculture
For Information on Water Supply Systems For Farm Homes

The dreary image evokes a sense of hard labor associated with getting water into our homes. In the upper part of the poster is the farm woman's purported dream: She is smiling, nicely dressed in short sleeves and colorful shoes, comfortably indoors, opening spigots at a sink, with hot water coming out, its steam curling to the ceiling. The poster is targeted at women, not men, a telling example of who endured the greatest burden from not having access to modernized systems, and who would reap the greatest benefit from modern improvements.

The burdened, vulnerable woman in the rural United States a century ago, struggling to fetch water, is not much different from the burdened, vulnerable woman in sub-Saharan Africa today.

For those women, their dream – the solution – is the same: a modern water system (with pipes and pumps) and a modern energy system, including electricity to drive the pumps and fuels to heat the water. Just as access to water and energy liberated that American farm woman, so it would for the billion-plus people around the world who do not have electricity or piped water today. Novel electrical appliances such as the dishwasher and washing machine provided women in the United States with even more freedom.

Water is difficult to pump by hand because it is so dense: It weighs 8.34 pounds per gallon. A typical person needs at least a few gallons per day for drinking. Add in a few more gallons for personal hygiene and yet again a few gallons for cooking and cleaning. The muscle power needed to raise that water out of a well or to crank the hand-pump is nontrivial. For a typical American, among the world's most profligate water users, typical usage in the home is 150 gallons per person per day because of watering our lawns, filling our pools, and our long hot showers, in addition to our basic needs.

Imagine pumping that water by hand, the way it once was. In poorer parts of the world that reality still reigns. In Southeast Asia and sub-Saharan Africa, it is usually the responsibility of women or girls to fetch the water. Those women and girls miss hours of work or school each day to get water from far away, carrying the heavy jugs of water balanced on their heads or hanging from bars resting on their shoulders, traipsing long distances to and in remote areas, which puts them at risk of sexual violence along the way.

Once they have the water, they are not done, as the water needs to be treated before it can be drunk. In remote villages where piped water systems and centralized water treatment with electrical pumps and other advanced techniques are not available, water is treated the old-fashioned way: It must be boiled. There's a similar story for getting fuel: Women often have to collect fuel from remote areas, and again they are vulnerable to assault along the way. In many parts of the world, women and girls spend an hour and a half per day collecting fuelwood, which again keeps them out of school. Using modern energy to spare girls the effort of those laborious chores gives them a chance for more school-based education.

When they return after having fetched water and fuels, they need to use the fuels to boil the water. Those fuels – including crop residues, animal waste such as cow dung, wood and untreated coal – are burned in stoves used for cooking and heating. Unfortunately, the dirty, inefficient cookstoves perform badly, producing indoor air pollution that has been linked to the premature death of about 2.5 million women and children every year. The International Energy Agency estimates that providing universal access to clean cooking systems would avert 1.8 million premature deaths annually by 2030. In other words, antiquated energy and water systems literally deprive girls of their education and kill women by the millions. Such archaic, labor-intensive and dirty approaches to energy and water take a toll on prosperity and economic opportunity.

In addition, the world's poorest women are also traditionally responsible for planting and harvesting crops, milling grain and fulfilling household chores as well. These responsibilities leave little time for an education or employment outside the home, which perpetuates poverty. Giving households in low-income regions like sub-Saharan Africa and Southeast Asia access to modern forms of energy, such as propane, piped natural gas and electricity, would also free up women and girls to attend school or earn extra household income. Educated girls have more choices, frequently marry later and opt to have fewer children, which helps to alleviate the cycle of extreme poverty.

Although the scenario I described sounds like something from the developing world somewhere far away, as the poster demonstrates, it is also part of the American experience in the not-too-distant past.

At the 100th anniversary of the promotional poster, it's fair to say that in the United States, the Farm Woman's Dream is now a universal reality. But since not everyone shares the benefits of energy and water access around the world, there is still much to do. If we invest in improving our energy and water systems, then that will benefit women, and therefore all of humanity, globally.

How Clean Energy Can Win Over Rural Areas

MECHANICAL ENGINEERING | FEBRUARY/MARCH 2021

> *Building out renewable energy infrastructure can bring opportunities to places eager for jobs.*

NOVEMBER 2020'S ELECTION RESULTS EXPOSED A LARGE RIFT between the politics of the cities and suburbs and those of rural areas. While voters in suburban Dallas, Atlanta, Indianapolis, and elsewhere moved to the left, many rural counties in across the country voted for President Trump in higher percentages than in 2016.

This growing separation in party politics affects views on issues such as climate change. According to a 2020 survey by the Nicholas Institute for Environmental Policy Solutions, rural voters are more likely to oppose government action to reduce climate change. They could throw up roadblocks to the kind of measures that are urgently needed to address this challenge.

But it is possible to craft a plan of action against climate change that bridges the rural-urban divide, bringing opportunities to places eager for jobs and economic growth while helping the nation meet decarbonization targets. We can do this by building large wind, solar and geothermal power plants everywhere we can–starting in rural areas.

The good news is that the best locations to build wind and solar farms and tap geothermal energy are predominantly in rural areas, away from the metropolitan cores. A concerted program to build a decarbonized power infrastructure would naturally start there. The first jobs would be in construction, but once these plants were established, they would need trained technicians and engineers to operate them. Opportunities for skilled positions and knowledge workers would help stanch the brain drain that so often affects rural areas.

The distance between these rural clean power plants and the metro areas that need the energy will require even more infrastructure: A new interstate electricity transmission system. But instead of installing ugly overhead powerlines that can spark wildfires or fall during storms, we could – and should – bury high-voltage direct current lines underground. Doing so is not only safer, but also more efficient, moves more power, and is less visually intrusive on the natural beauty of these areas.

What's more, underground lines can take advantage of the existing rights of way of interstate highways and cross-country railroads to expedite permitting and avoid the need to traverse private land. Analyses by national labs and industry show that such a national network of power lines and renewable sources would reduce costs and emissions while improving reliability.

These clean-energy power plants and long-distance electricity transmission lines would support the rural economy. When electricity flows from rural areas to cities, money flows from the cities back to the countryside to pay for the electricity. Wind, solar and geothermal power can become supplemental cash crops alongside traditional agriculture.

Clean energy can provide jobs even for those who today work in fossil fuel industries. Petroleum workers have the expertise in geology and drilling needed to tap subsurface geothermal heat that can be a valuable supplement to wind and solar power. And coal miners have the skill set to reclaim old surface mines, restoring them to wooded landscapes capable of scrubbing CO_2 out of the atmosphere as trees and other foliage grow back.

The impact of a move toward decarbonized energy can support rural economies in other ways. Policies such as renewable portfolio standards, R&D at national labs and research universities, tax credits, workforce development, rural renewable economic opportunity zones, infrastructure money and other incentives can speed up the process of cleaning up the air, establishing a technological lead, supporting domestic technologies and fuels, and making our grid more resilient all while injecting economic vigor in depressed areas. It's the kind of suite of policies that rural Republicans (who want economic growth, rural revitalization and a place for fossil fuel industries) and urban Democrats (who want the low-carbon, renewable technologies installed) can agree on.

Rapidly expanding a national grid and connecting rural renewables at one end and urban customers at the other just might interconnect us in economically vibrant ways that help temper the political divide while also staving off the worst effects of climate change.

Learning How to Beat the Heat

MECHANICAL ENGINEERING, AUGUST 2019

As the world gets hotter, places will need to adapt their infrastructure – and long-held cultural traits.

IF YOU'RE FROM TEXAS LIKE I AM, YOU ARE ACCUSTOMED TO HOT weather. But the epic heatwave that descended on Europe in late June 2019 was a whole other beast. Wildfires broke out in Spain, spontaneously igniting – among other things – farmers' manure piles. Towns in France hit 45°C (113°F), breaking heat records for the instrumental age.

Even though I'm used to heatwaves, hot summers and scorching sidewalks, a 105°F day in Austin, Texas, is easier to take than a 95°F day in Paris, where I'm writing this column. In Texas, the ubiquity of air conditioning, ice, swimming pools, shade and screens on windows means that hot days there are manageable. In France, when it gets this hot, there is no escaping the heat.

Because Paris is in a northern location with a cooler climate, Parisians design their buildings to retain heat rather than shed it. The buildings don't allow airflow that would let heat escape and thus are more energy efficient in winter. What's more, limestone, granite and other rocks are typical building materials – one of the reasons Paris is considered one of the most beautiful cities in the world. But lack of ventilation traps heat in buildings in hot weather, and stone holds a lot of heat and reflects a lot of light, exacerbating the urban heat island effect.

It isn't just infrastructure. France does not have an "ice" culture the way Texas does. In Texas, ice is served in water, tea, sodas and most other drinks. Beer is ice cold. Cold-brewed and iced coffees are common. In France, restaurants won't put ice in water unless it's requested, and even then it is usually just a few cubes of ice for a glass of water.

Infrastructure and culture can interact to stymie common cooling solutions. Take windows: The classic French window is tall with glass panes that can either swing open or be covered by working shutters.

(Most shutters in the U.S. are merely decorative.) Such windows work well for letting in fresh air or defending against prowlers, but they make installing window air conditioning units awkward or impossible. An alternative – mini-split systems that place the compressor on the roof and pipe in chilled refrigerant through the walls – has trouble cooling large spaces and can be difficult to retrofit into the older building stock typical of Paris.

As a result, home air conditioning in France lags the United States. That difference can have deadly consequences. When a heat wave hit Europe in 2003, up to 70,000 people died across the continent, including more than 15,000 (mostly elderly) people in France. And cultural idiosyncrasies played a major role: The heatwave hit in August, the traditional vacation month in France, so fewer doctors were on call. Elderly people, not used to intense summer heat, didn't have fans and didn't know to stay hydrated.

Compounding things, many elderly lived on fixed incomes and could only afford top-floor apartments in walk-up buildings. Those apartments are hotter than the average home.

To avoid such a disaster during the 2019 heatwave, public officials waived entrance fees for swimming pools and closed un-air-conditioned schools at midday so kids could cool off at home. They also recommended people go to movies or shopping malls. (This was a big selling point for American cinemas before home air conditioning was common.) But the shopping malls in France have weak ACs and were crammed with people – who went there to cool off – so they were hot and stuffy. My family went to a modern mall one early evening during the heat wave, and it was hotter inside than outside.

All of these factors reveal the complexity of climate change and the need for localized solutions, as the challenges will vary. Labor, medical and housing policies will play an important part. Some of the infrastructural concepts common in hot climates, such as shade trees and better ventilated buildings as well as more widespread air conditioning, will have to be adopted by places that have up to now been more focused on fending off winter chill.

Most importantly, one key takeaway is that if rich, developed countries in northern latitudes have trouble adapting to climate change, just imagine how tough it will be for poorer, hotter regions as they get even hotter.

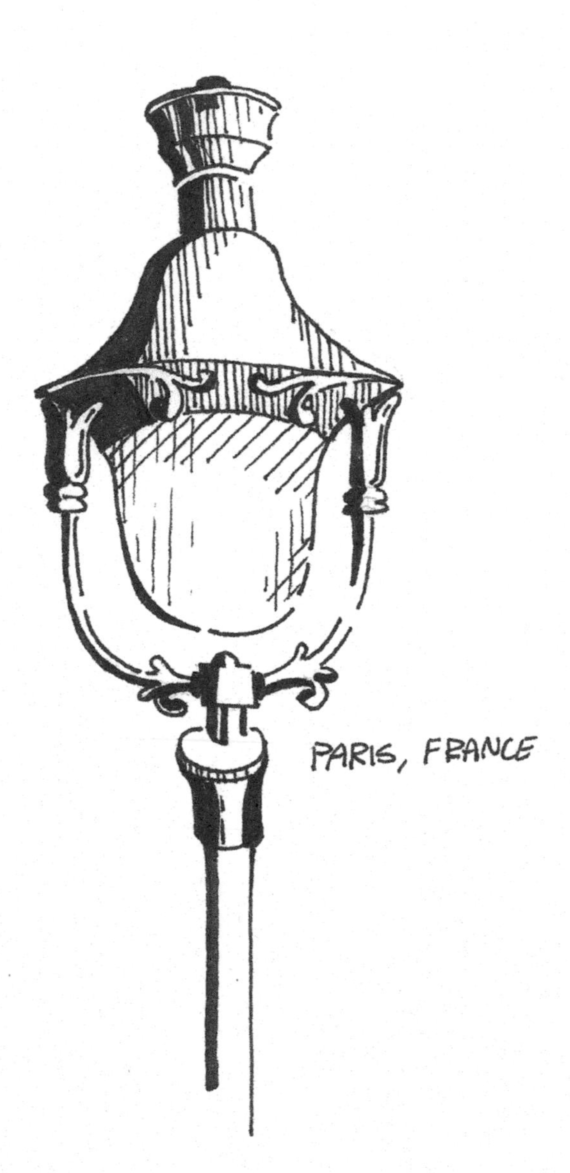
PARIS, FRANCE

Chapter 10

Self-Actualization Needs II: Energy & Culture

Energy Industry Needs a Better Approach to Communication

HOUSTON CHRONICLE, MARCH 2015

> *The energy industry has become that out-of-touch monolithic institution that doesn't seem to understand what it's being asked.*

DURING THE 1990S, WHEN MICROSOFT WAS WIDELY LAMBASTED for being wonky and out of touch with customers, a famous joke went like this: "A man riding in a hot air balloon gets lost and yells down to the first person he sees, 'Where am I?' 'One hundred feet up in the air,' says the man on the ground. 'By chance do you work for Microsoft's technical support?' says the man in the balloon, 'because your answer is technically correct but completely unhelpful.'"

Microsoft spent more than a decade working to repair that reputation. The energy industry needs to learn the same lesson.

Today, with contentious energy questions front and center about the safety of exploding oil trains, climate change, fluctuating oil prices, methane leaks, risks of water contamination and earthquakes, the energy industry has become that out-of-touch monolithic institution that doesn't seem to understand what it's being asked.

Consider the typical energy industry response to questions about hydraulic fracturing, or fracking.

It is not unusual at town hall meetings for concerned residents of regions with rampant shale production to ask oil and gas company representatives whether the fracking is causing water contamination and earthquakes. Industry representatives will often triumphantly respond with great confidence and clarity that there is no evidence that fracking has contaminated aquifers or caused earthquakes. These answers are technically correct but not helpful at all.

To the average person, fracking is not just a specific technical process. It is also the word that represents the whole chain of events for producing shale oil and gas, including the trucking, drilling, noise and dust. Concerned residents can therefore be excused for making the logical assumption that induced earthquakes and water contamination are associated with "fracking." Like the man in the balloon, the public sees the lay of the land on these issues more clearly than the energy representative who responds to their concerns with narrow, technical answers.

It's true that people could ask better questions. The man in the balloon should have asked, "How do I get home?" instead of "Where am I?" Concerned citizens should ask, "How do we ensure the quality of our water and reduce the frequency of earthquakes?" instead of asking, "Is fracking the cause?"

But, as Microsoft learned, corporate leaders can't afford to wait for people to ask the right questions. They need to meet the public halfway and recognize the obvious facts: The man in the balloon is lost, and the residents of shale-producing regions are worried.

I often work with representatives from the energy industry, and I believe they are not as out of touch and dismissive as they appear. They are just answering the questions they were asked, instead of the question they should have been asked.

Maybe they just aren't listening carefully, a problem that can afflict any of us. But occasionally, their vested self-interest means they don't want to or can't tell the entire truth. And, regretfully, some industry representatives know the right answer but don't feel empowered to deliver it.

To the people who are concerned, all three look the same, so it's hard to distinguish between the clumsy communicator and the person who has something to hide.

Unknowingly, people in the energy industry often look like they are trying to get away with a technicality rather than coming clean on the actual risks. Instead of hiding the details, make the case that there is good news and bad news with energy, but done the right way, the benefits outweigh the risks. Instead of dismissing concerns and speaking down to people, acknowledge the concerns and spend more time, money and effort on outreach, communications and energy literacy.

It doesn't matter how sophisticated your science is if your local stakeholders won't let you operate because they don't trust you. But regardless of how people phrase their questions, the energy industry should listen carefully for the real questions being asked. We can't just remind the man in the balloon that he's lost. We must give him directions.

How "Frankenstein" Prevents Us From Tackling Climate Change

EARTH MAGAZINE, JANUARY 2016

By Michael E. Webber and Sheril R. Kirshenbaum

> *The apocalyptic conditions after the eruption of Tambora subsided in about three years, but this incidence of temporary naturally occurring climate change left a surprising and persistent cultural impact, which has influenced our current reticence to deal with anthropogenic climate change.*

JUST OVER 200 YEARS AGO ON THE INDONESIAN ISLAND OF Sumbawa, Mount Tambora erupted with an explosive plume that ejected debris up to 43 kilometers into the atmosphere, darkening daytime skies. Molten rock and pyroclastic flows cascaded down the flanks as ash spread out over hundreds of thousands of square kilometers. This was the most powerful volcanic eruption of the past 500 years. Tsunamis surged. Tens of thousands of people died directly from the eruption, with hundreds of thousands more dying in its aftermath from the starvation and disease that followed widespread crop failures.

Scientists have documented how this event triggered global cooling: When sulfur released from the volcano reached the stratosphere, it spread around the equator and toward the poles. Once oxidized, tiny reflective particles limited the amount of sunlight reaching Earth's surface and the entire planet got colder. Average temperatures temporarily dropped approximately 1°F. As a result, the year 1816 became known as "The Year Without a Summer."

The apocalyptic conditions subsided in about three years, but this incidence of temporary naturally occurring climate change left a surprising and persistent cultural impact, which has influenced our current reticence to deal with anthropogenic climate change.

During the unusually dark and stormy summer months of 1816, British teenager Mary Wollstonecraft Godwin (better known as Mary Shelley) and her poet friends, including Lord Byron, were forced to stay inside their chalet at Lake Geneva to avoid perpetual rain. To pass the time, they read ghost stories and eventually agreed to try writing their own. This spurred Shelley to compose her now-classic masterpiece *Frankenstein.* The story gripped the public's imagination, but also unintentionally shaped their fears of science, fostering distrust of scientists and the scientific method.

The 1931 movie interpretation of Shelley's book seemed much like a treatise against science. The film focused on the theme of scientists trying to play God, which led to catastrophic consequences. The monster that scientist Dr. Frankenstein unleashed on society left a lasting impact. Famed science writer Stephen Jay Gould lamented the influence of *Frankenstein* on popular culture, noting how subsequent movies have reinforced the idea that scientists like to play God and put society at risk, and so should not be trusted. *The China Syndrome, 12 Monkeys, Austin Powers, Spiderman* and *Jurassic Park* are other examples over past decades that reinforced this theme. There are many, many more.

Unfortunately, the cumulative impact of films like these coupled with politically motivated misinformation campaigns about science has become a barrier to using science to inform public policymaking and has fostered a widespread suspicion of the scientific community. This mistrust plays out in a variety of ways, from frequent accusations of scientific arrogance to complete dismissal of climate change altogether.

So, in a surprising fate of circumstances, naturally occurring climate change – from a volcano – spurred an imaginative story that, in turn, made society wary of scientists. Now two centuries later, human-induced climate change – from the combustion of fossil fuels, agriculture, and land-use changes – has made our world increasingly vulnerable to flooding, extreme weather events, drought and instability. Unfortunately, a deep-rooted skepticism toward science and scientists persists, largely along political lines and over certain issues, which means scientists' dire warnings about climate impacts go unheeded. As a result, we are slow to act on one of the gravest risks to modern civilization.

Now, as we need our scientists the most, they are trusted the least – at least when it comes to climate change, thanks to blatant misinformation campaigns. It is time for us to get over our skepticism and lean on the science community for fact-based insights on the world around us, without interference from politics, misinformation campaigns, or even dramatic license in Hollywood and literature. If we do this the right way, then we can avoid another more-magnified scenario of the dire consequences the 1815 event kicked off.

Unlike the sulfur particles emitted by Tambora, the causes of anthropogenic climate change are not going to dissipate – nor will their impacts abate – in a matter of a few years.

And if we act fast, it won't be too late.

A Pitch to Study BREW: The Beer-Renewable Energy-Water Nexus

EARTH MAGAZINE, SEPTEMBER 2017

> *Like the food-energy-water relationships, the relationships among beer, renewable energy and water have been strong throughout history, and until very recently, the demand for all three was increasing. Today, they are headed in different directions. We should study this.*

OVER THE PAST FEW YEARS, THE FOOD-ENERGY-WATER NEXUS[55] has become trendy for multidisciplinary systems-level research as it ties together three of the most critical elements of a modern society. Food security, energy security and water security are all inextricably linked, and the demand for all three is increasing. In recent years, the U.S. National Science Foundation has set aside $75 million to support research that spans these three topics. While I applaud this research path, I would like to suggest another topic that also warrants dedication from researchers: the beer-renewable energy-water nexus, or BREW.

55 See Chapter 1

And I would further like to be the first to volunteer to personally collect field data at pubs and breweries nationwide.

Like the food-energy-water relationships, the relationships among beer, renewable energy and water have been strong throughout history, and until very recently, the demand for all three was increasing. Today, they are headed in different directions. While the demand for renewable energy and water has never been higher, beer consumption has begun falling. Meanwhile, energy sources are becoming more distributed, as are beer producers, and we face increasing threats to our water supply (and, thus, beer). Clearly, this is a field ripe for study.

Changing Energy Sources

In the late 19th century, around the time of the second Industrial Revolution, the world transitioned to using coal as a major energy source, but beer producers made that transition nearly two centuries prior. Coal was a lower-cost, higher-performing fuel, producing more heat and less ash and smoke than wood. Coal's greater energy density made it perfect for steel-making, improving the quality, consistency and robustness of the metal. But coal, or specifically, coke – coal first baked in ovens to remove impurities, which burns hotter and cleaner – also improved the quality of beer. The introduction of coke into beer-making in the early 18th century enabled production of a higher and more consistent quality of beer favored by wealthier consumers.[56]

Coal transformed society. Out was the variability of renewables (wood in this case) and in was the consistent quality that urban affluence demanded. But that also led to industrial consolidation worldwide, as smaller companies were absorbed into larger multinational conglomerates. The same trend happened with beer.

As fossil fuels became widely available, coal- then diesel-fired rail and truck transportation with refrigerated cargo holds made the

56 When coke from coal was used to roast the malts, they were less dark than if wood had been used to roast the malt, which gave rise to the term "pale ales."

movement of beer much easier. Consequently, a wave of consolidation unfolded for a century, with the number of brewing facilities reaching a low point in 1978 when the United States had fewer than 100.

While fossil fuels were liberating for society as a whole, they changed the beer industry. We went from small, local breweries – originally designed to serve local markets because the lack of refrigeration and slow transportation made it hard to ship beers to far-flung markets without spoilage – to remote, massive, centralized beer factories that gave us reliable distribution, but of a bland, easily digestible liquid.

But that trend is reversing nowadays, for both energy and beer. Just as the energy sector is moving from massive centralized, corporatized (largely coal-fired) power plants toward distributed local systems with more variable power sources, such as wind and solar, many consumers are demanding the same thing for beer, pushing for distributed craft production of high-quality brews with a variety of unique local flavors and styles, including seasonals. Today, there is a boom in microbreweries, craft beers and beer festivals. At the end of 2016, there were about 5,300 breweries[57] operating nationwide, breaking the long-standing record set in 1873.

Biofuels and Beer

Biofuels and beer also have a historical connection, although today it is more adversarial. Corn-based ethanol was among the first major fuels used for the internal combustion engine. Ford's famous Model T worked on ethanol from corn. Car designers at major automotive companies like working with alcohol-based fuels – they used to call it gasohol – because they allow for higher compression ratios, which enables higher performance, which is especially useful for racing. Some race cars still rely on a fuel made from 98 percent ethanol. Alcohol is still the primary renewable fuel used for transportation in the world today. It's almost as if we're putting beer into the gasoline tanks of our cars.

57 There were about 8,500 breweries in 2019, so the trend continued.

Today, beer and fuel ethanol production are in direct competition, as the two industries (along with agriculture) jockey for limited supplies of grain. This competition is what led breweries to fight against the national biofuels policy, which in the Energy Independence and Security Act of 2007 required that 36 billion gallons of liquid transportation fuels (about 20 percent of that year's consumption for transportation) come from renewable sources. The act was appealing because it might have improved energy security by reducing imports of petroleum. But a byproduct of the biofuels policy was a rippling effect on commodity prices throughout the agricultural sector, which raised, among other things, the cost of malt barley, a key ingredient in beer.

Water and Beer

Another key ingredient in beer, perhaps the most important one, is water. Historically, brewing was a way to purify water, or at least to get your daily liquids in a safe way. Ancient people discovered that the fermentation and distillation process killed pathogens, providing a sanitary, although intoxicating, means of staying hydrated. One suspected reason why monks were not afflicted with cholera at the same rates as other groups is because they only drank beer they brewed on their own.

Today, the role of beer as a protector of water quality plays out a little differently.

In 2010, Enbridge Inc., a company operating in the Canadian oil sands, spilled diluted bitumen from a pipeline into Michigan's Kalamazoo River, from which a local brewery, Bell's, drew its water. Enbridge Inc. was ordered by the Environmental Protection Agency to initiate a massive dredging program to scrape the pollution from the sand and soils at the bottom of the river. Due to concern about what other pollutants the dredging might release, Bell's Brewery sued to halt the dredging near brewery operations. "When energy's impact on water starts screwing with my beer...that's it, man. That's where we draw the line," noted a writer in a July 2013 Daily Kos article on the case.

That's not the only example of a brewery working to protect local waterways. The Red Hills Brewery in Alabama works with conservation groups and scientists at Auburn University to protect water quality on behalf of the brewery's adopted mascot, the native salamander, which needs moisture on its skin to breathe. Many other breweries across the country are also involved in water protection.

In fact, the Natural Resources Defense Council even has a program called Brewers for Clean Water.

A New Discipline?

Given this collision of national policies related to alcohol consumption, energy, the environment, water, national security and agriculture, there is clearly much to study. Luckily, few things universally interest scientists and engineers more than beer. We like to drink it. We like to talk about it. Some of us like to make it. Sometimes we even like to research it and write about it.

World War G: Zombies, Energy and the Geosciences

EARTH MAGAZINE, DECEMBER 2013

Spoiler alert: If you haven't read or watched World War Z, *but you intend to, note that this column contains spoilers.*

> *Energy and climate change issues weave in and out of* World War Z. *While the zombie element is fun, the book ends up being a serious critique of modern civilization's inability to swiftly and firmly deal with a major global crisis that puts the fate of societies at risk.*

WHAT CAN WE LEARN FROM PEOPLE WHO HAVE SURVIVED A ZOMBIE INVASION? A LOT, IF YOU FOLLOW THE BOOK (AND MOVIE) *World War Z* to a logical conclusion. Although the book and the movie are very different from each other – they share the same title and topic (zombies), but that's about it – energy and climate change issues weave in and out of both stories.

The book is presented as an after-the-fact oral history that reveals some key elements of a supposed war waged by humans against a zombie invasion. Although humans were ultimately successful, victory was achieved at great cost. Nominally about the rise of the undead, *World War Z* includes gripping details about how the zombie infestation spread, the terror it induced, and the lengths to which people went to survive.

While the zombie element is fun, the book ends up being a serious critique of modern civilization's inability to swiftly and firmly deal with a major global crisis that puts the fate of societies at risk. In that way, it reminds me of how we're dealing with climate change. Or, rather, how we're *not* dealing with climate change. In fact, the movie captures this stark similarity in the very opening scene, where the narrators (in the form of news anchors) discuss the threat of climate change interspersed with breaking updates about the "rabies" outbreak that is later revealed to be the zombie attacks.

In the story, the governments of the world pretend nothing is wrong so as not to worry the populace. The markets sell snake-oil solutions, profiting off fear while giving the governments and people a false sense of security. Because of the delays in responding, the problem grows worse than expected and harder to solve, taking a significant toll on human life and prosperity along the way.

The fictional zombie invasion in *World War Z* can be interpreted as a symbolic stand-in for real-life climate change. In fact, while reading the book, I couldn't help but replace the words "zombie invasion" with "climate change," "pandemic[58]," "global economic collapse," and any of a number of other multicontinent crises. The storyline would be largely the same: failure to recognize the global threat for what it is, profit-seeking industrialists who confuse real solutions with self-serving

58 Unfortunately, the COVID-19 pandemic affirmed the thesis of the book. Early commentators referred to the outbreak of COVID-19 as a mysterious flu, reminiscent of how the book's commentators mistook the rise of the undead as a form of rabies.

falsities, acting too late, failure to adjust our approach because of a short-sighted sense that things surely aren't all that bad, and then wondering how it all went wrong. Hopefully that's not the path we will take with climate change.

The connections don't end there. The story is laced with details of energy concerns that are analogous to real life in some surprising and some not surprising ways. Notably, energy systems are the indicators of death or survival: The cities that collapsed had lost their central power grids, and the cities that survived were able to protect their energy systems. The story subtly but repeatedly notes that survivors used independent, off-the-grid energy systems fueled by biomass or solar energy. The centralized systems that depended on fossil fuels, large powerplants, transmission lines, refineries and pipelines failed.

Whether consciously or not, the author is reminding us that the self-reliance ethic that has been central in more than two centuries of American philosophy is compatible with small, distributed, renewable energy systems rather than large, centralized, fossil-fueled systems. He seems to be saying that the key to survival against zombies or climate change is the marriage between two extremes of modern American political thought: survivalists with an off-the-grid mentality, who want democratized and locally controlled energy systems, and environmentalists who want renewable and low-carbon options. Technologists get tossed a bone too, as a small nuclear system embedded in a submarine makes a cameo as the savior for one community, giving them electricity and clean water.

Bringing the whole climate change analogy full circle, the zombie infestation actually causes climate change in the world of *World War Z*, and one of the survival methods is to move toward the poles. It turns out that zombies freeze in the cold, so moving to cooler climates proves the best option for adaptation to the new reality. Survivors set up refugee camps and "free zones" (meaning zones that are free of zombies) in Nova Scotia, for example. Simply moving toward the poles instead of doing the hard work to prevent the problem from growing in the first place is the preferred option. Hmm. This sounds eerily familiar, as some policymakers are suggesting we simply move to colder climates instead of mitigating greenhouse gas emissions.

In reality, I hate to think that we would need to turn to a book about a fictional zombie war for guidance on climate policy. Then again, it may hold some valuable insights for us, considering how little we've been doing for the last few decades.

Energy and the Super Bowl

KUT RADIO STATION (AUSTIN'S NPR AFFILIATE), FEBRUARY 3, 2014

> *About 110 million people watch the Super Bowl on Sunday in about 40 million households. That means across the nation we needed 4 gigawatts of power to watch the game. That's the power output from about four nuclear power plants or two Hoover Dams. And that's just for the TVs, not including the lights, heating, cooking or anything else.*

THE SUPER BOWL IS AN ENERGY HOG. IT'S NO SURPRISE THAT AN arena with more than 80,000 people attending and with more than 100 million people watching at home would require a lot of energy. At the stadium there are the lights, the heating, and the massive jumbotron.

The head of the local utility that provides electricity to the Super Bowl estimated that it required 18 megawatts to power the entire complex for the game. That is about the amount of power needed from a jet engine of a Boeing 747 running at full throttle. That kind of demand can really strain the local distribution grid for electricity. It is no wonder that [the 2013] Super Bowl had a blackout lasting 30 minutes.

At home, we use a lot of energy for our televisions to watch the game. Just a few years ago, TVs required about 300 watts apiece, which makes them one of the most energy-intensive appliances in the house. Because of efficiency improvements, new TVs are much better and only require 100 watts, which makes them only as bad as the worst old-fashioned lightbulbs.

There were about 110 million people watching the game on Sunday in about 40 million households. That means across the nation we needed 4 gigawatts of power to watch the game.

That's the power output from about four nuclear power plants or two Hoover Dams. And that's just for the TVs, not including the lights, heating, cooking or anything else.

While the TVs were used presumably to watch the game, as always, the real competition was in the commercials. I was particularly interested in how energy might show up in the commercials. In prior years there have been the feel-good ads calling for energy efficiency, and who could forget Audi's "green police" ad from 2010. The car ads are particularly important, as transportation is responsible for 28 percent of U.S. energy consumption each year.

The Toyota Highlander commercial caught my eye for two reasons: It starts off by passing an oil pumpjack in a barren landscape during its opening seconds, a subtle nod toward its fuel economy. And, it features Kermit the Frog and the other Muppets, which actually makes a lot of sense since the latest Muppets Movie was all about how an evil oilman named Tex Richman was trying to bankrupt them so he could drill for oil at the old Muppet theater.

I had to wonder if Toyota simply forgot that Kermit the Frog was used as a pitchman in 2006 for the Ford Escape Hybrid, concluding that because of the car's fuel efficiency, "it is easy being green" after all. A VW commercial showed German engineers getting wings, which made me wonder how much oil we could save if we used them instead of planes for flying. The all-electric Smart Car's ad was clever. It compares the silliness of using a golf-cart sized electric car to go offroading with the silliness of driving a large SUV in the city. I have to admit that its advertised price of $139/month sounds like a pretty good deal, especially if it means not paying for gasoline ever again.

But my favorite ad was Duracell battery's "the power within" ad. It's about the amazing story of Derrick Coleman, who is deaf but overcame great challenges to make it to the NFL. For me, it's fitting that a battery company would sponsor this ad, not only because batteries in hearing aids are an important tool for the hearing impaired, but because on a global scale, for the many millions of people around the world who are in poverty, a little bit of energy goes a long way toward lifting us toward a better life.

Even at the Super Bowl, there are energy lessons for all of us.

Energy and Christmas: Not Just a Lump of Coal in Our Stockings

UNPUBLISHED, WRITTEN DECEMBER 2015

> *Christmas may well be our peak energy event of the year.*

HOW ARE CHRISTMAS AND ENERGY INTERTWINED? WE DON'T often think of them together, but Christmas may well be our peak energy event of the year.

For starters, all the toys and games under our trees are made from energy; the plastic ones are derived from petroleum. As a nation, 5 percent of our energy consumption is physically embedded in the things we make: Petroleum or natural gas liquids go into a chemical plant, and plastics come out that are subsequently fabricated into toys in factories often powered by coal.

Then there's the energy that goes into our celebrations. Not just the food and the traveling, but also the shining lights of Menorahs and the Christmas trees. To bring light to the dark of winter, we go all out burning and consuming energy at this time of year. Ages ago our ancestors lit Yule logs, oil lamps and candles. In honor of the Yule log, we may still light a ceremonial fire, or just dial up the central heating a notch or two. Today we turn on the Christmas lights—hopefully new and efficient LEDs, but often not. Christmas movies often make a mockery of ambitious Christmas light displays (ahem, Clark Griswold) that threaten to take down the whole grid.

Even the millions of tablets and smartphones that will be opened this December have embedded energy. The smartphone was only made possible by the sleek, portable energy provided by the lithium-ion battery inside it. No battery, no cellphone. For that matter, the whole Internet depends on electricity. Without ubiquitous power, the information economy would never exist. So when you watch a movie on your tablet this month, remember, you're also tapping into a sophisticated grid combining energy and information.

Each year seems to give us smartphones that are more efficient and powerful. We need to expect the same progress in the energy that powers these gadgets, as we move further away from Yule logs and candles.

Consider the lessons of the lump of coal. For many kids even today, there is the looming threat that if they are naughty, the gift from Santa Claus in their stockings will be a lump of coal instead of a shiny new toy. That this is considered a punishment rather than a reward shows how far we have come. In the 1800s and earlier, a lump of coal was a desirable gift. Coal was expensive and had remarkable energy density. A gift of coal meant heat in the middle of a cold winter. It's only since the advent of furnaces powered by cleaner, more-efficient forms of energy like propane, natural gas and electricity that a lump of coal became undesirable.

That transition is a reminder of how much energy can change not just our quality of life, but our mindset too. Holiday traditions can change because of what energy can do for us. And as we celebrate our holidays together, it's worth taking a moment to recognize and enjoy the energy surrounding us. Our energy problem is a complicated one, so we need compromise, collaboration and continued innovations for a better future for energy and all that enjoy it.

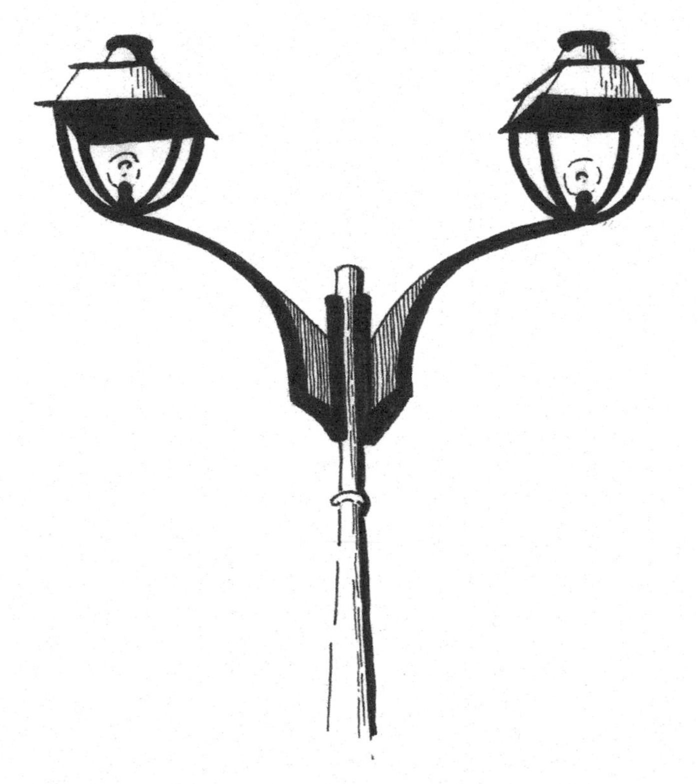

LOIRE VALLEY, FRANCE

Chapter 11

Self-Actualization Needs III: The Folly of Predictions

Throughout 17 years of authoring essays, I have made many predictions about the future of energy. Enough time has passed that it is possible to go back and review what I got right and what I got wrong. This chapter includes a sampling of my musings, including some that were prescient and some that were flat out incorrect. I've updated a few items here and there in these essays below and also added some commentary with self-reflection about their accuracy. Enjoy!

The Risky Business of Predicting the Future

MECHANICAL ENGINEERING, DECEMBER 2018

Mistaken forecasts can limit our progress.

FRANK WHITTLE WAS A BRITISH ENGINEER WHO (NEARLY simultaneously with Hans von Ohain in Germany) invented the first turbine-based jet engine, successfully demonstrating it in a test flight on

May 15, 1941. Whittle's designs – including the incorporation of an afterburner to give more thrust – are still the basis for powering most major civilian airplanes and fighter jets. The gas turbine is without a doubt one of the most important inventions of the 20th century.

A year earlier, however, the National Academy of Sciences published a treatise by its Committee on Gas Turbines concluding that harnessing the gas turbine for aeronautic power was "beyond the realm of possibility with existing materials."

At some point, Whittle received a copy of the report. Directly under that passage, he wrote in the margins, "Good thing I was too stupid to know this." That anecdote reminds me of Clarke's First Law – Arthur C. Clarke's wry comment on the limits of expertise: When a distinguished but elderly scientist states that something is possible, he is almost certainly right. When he states that something is impossible, he is very probably wrong.

Clarke wrote that in his nonfiction book, *Profiles of the Future*, an exploration on the way scientists and other would-be prognosticators misread present-day trends and make pronouncements that often wind up sounding silly in retrospect. Some of these predictions are justifiably famous – for instance, that heavier than-air flight is impossible, or that passengers on a fast-moving train would end up suffocating. Other predictions perhaps ought to be famous (or infamous), such as Digital Equipment Corporation founder Ken Olsen's statement that there is no reason why anyone would want a computer in their home.

It would be easy to simply shrug off these failed predictions, but they affect our decision-making today, which can inhibit desirable pathways into the future. For some sectors, much of what we will be able to do in the future depends on our ability to distinguish between what is possible and what's preposterous. This is especially true for the energy industry, which relies on expensive and long-lasting infrastructure. Build the wrong facility in the wrong place, and we will be stuck with that mistake for decades.

Famed energy writer Vaclav Smil has also struggled with the difficulties of our energy future and complained openly about it.

"Forecasters often merely extrapolate existing trends, unreasonably assuming that the underlying conditions will remain stable," Smil has written. "Thus, pessimists and environmentalists commonly see doom around the corner, whereas technophiles and optimists often envisage a coming paradise."

A case in point is wind and solar power: Authoritative forecasts over the last decade from the U.S. Energy Information Administration and the International Energy Agency have consistently overpredicted their costs and underpredicted growth in their adoption. Some people of a conspiratorial bent might see the undue influence of the fossil fuel industry, but the probable cause of those missed forecasts is seeing the future as a simple extension of the recent past and failing to anticipate the evolution of technology or the learning gained from ramped-up production.

Those bad predictions have the net effect of inhibiting investment in renewables and setting us back from the progress we need to make. Had policymakers realized just how inexpensive power from wind turbines and solar panels could be in 2018 – for many applications, wind and solar are just about the cheapest new energy sources around – they might have not only held off on adding more fossil fuel-powered generating capacity in recent years, but also invested heavily in technology such as energy storage and smart grids that enable greater use of intermittent power sources.

As I approach 50, I can see from the look in my students' eyes that I am closer to being one of Clarke's "distinguished but elderly scientists" than I care to admit. It's important to support the next generation of energy futurists, who are blinkered neither by expertise or incrementalism and can think boldly and broadly about what is coming next and what should happen. Until we do, we will continue to stumble along.

A Fool's Look Into the Future

EARTH MAGAZINE, DECEMBER 2009

> *Any good observer of humankind's relationship with energy would recognize the folly of making predictions about the future of energy.*

IN THE 1985 CLASSIC *BACK TO THE FUTURE*, DOC BROWN, the mad scientist played by Christopher Lloyd, queries his time-traveling visitor, Marty McFly (Michael J. Fox), in 1955 about the future of America with the simple question of who was president. The answer of Ronald Reagan appeared astounding and ridiculous, as his name was on a movie poster at the time.

Such is the way it is with an innovative and evolving society such as ours: Ideas and concepts that seem ridiculous in one era make sense in another. At the end of the movie, Lloyd returns to 1985 from a more distant future with a coffee grinder-sized "Mr. Fusion Home Energy Reactor," which he fuels up with trash to provide the 1.21 gigawatts of power he needs to operate his DeLorean's time-travel-enabling flux capacitor. With a twist of irony, perhaps the movie's predictions of handheld nuclear devices will prove to be startlingly accurate. The recent announcement of Hyperion's hot tub-sized, sealed, distributed nuclear devices that can supply enough power for 20,000 average U.S. households suggests we're further along that path than one would have expected.

It is within this context that I'll lay out some of my thoughts about what our future energy systems hold for us. Any good observer of humankind's relationship with energy would recognize the folly of making predictions about the future of energy. After all, in the energy world, it is just as easy to make predictions that are correct for all the wrong reasons (for example, because of the odd fortune of including two mistaken assumptions that cancel each other out), as it is to be wrong for all the right reasons. And thus it is only a fool who would be bold enough to put pen to paper to mark the permanent written record with his or her predictions about energy. I am that fool. And these are my predictions.

Oil Prices Will Go Up ... A Lot ... Before They Crash

As society struggles to find ways to displace oil, we will continually bump up against the fact that for transportation, it's simply the best fuel on the market. That's hard to compete with and will introduce delays into our energy transition.

At the same time, we have two parallel trends: developing societies that want more oil than ever before and the increasing cost of producing oil because of aboveground reasons (developed societies that don't want oil and are implementing policies that make it more expensive to find, produce and consume petroleum) and belowground reasons (the most accessible oil has already been produced). These countertrends mean that demand for petroleum will stay high at the same time access to its resources will be more expensive. These will combine to push prices higher until the time that an abundant, sustainable and economical alternative fuel finally arrives, likely two to four decades from now, at which time demand for petroleum will plummet and prices will crash. That is, petroleum might follow the path of whale oil, whose prices collapsed when something better – namely petroleum – came along.

> *Postscript: I'm not sure how to judge this one. In December 2009, oil prices were quite low as the world was in economic recession. Thus, no surprise, prices indeed rose significantly. Then fell again, rose again, fell again, rose again, and so forth. But global oil prices didn't reach the highs witnessed in 2008 until 2022, when the Ukraine-Russia war spiked prices. I got the key points correct that oil will remain critical for a while and that harder-to-reach petroleum will provide a bigger slice of our oil, but I failed to anticipate the technological improvements that would make shale production cheaper as time went on. I do think the fate of petroleum could follow the arc of whale oil, but basically missed it on prices.*

Natural Gas Prices Will Go Up, Bringing Abundance

Could natural gas be to oil what oil was to whale oil? Maybe. Natural gas prices have been quite low recently, but as a part of a general upward trend in energy prices that we're going to see, natural gas prices will climb back up, and in the process, the United States will (re)discover its natural gas. Then we'll see a lot of changes.

> *Postscript: Nailed it. The shale revolution unleashed a torrent of natural gas in the United States, lowering prices, flipping liquefied natural gas imports to exports, and helping to displace coal in the power sector.*

U.S. Congress will discover that natural gas is an abundant, clean, domestic resource, something that apparently slipped its mind during the development of the American Clean Energy and Security Act, which the House passed in 2008. It's as if the entire bill was written about coal and oil, and its authors forgot that natural gas is the primary source for one-fourth of our energy consumption, is mostly produced domestically, has half the carbon footprint of coal (or better, when used with combined cycles), one-third lower carbon emissions than petroleum, and is a part of a domestic resource base that grows with time. For electricity, we can use natural gas to displace coal and to supplement our variable wind and solar resources – in the process improving our air quality, security and economy. Once Congress realizes that we have a lot of natural gas and that natural gas is a great enabler of renewable electricity, many of our energy problems will be solved for at least a few years.

> *Postscript: Missed it. Congress has not embraced natural gas and still prioritizes coal, oil, nuclear and renewables in its policymaking.*

Furthermore, to displace oil, we could use natural gas to power up our cars. In the near future, Americans will discover that natural gas vehicles – already in existence – are cost-competitive with hybrids, plug-in hybrids and diesels. All this talk about electric cars is fun, but is also part of a century-old competition between batteries and gasoline tanks. The dark horse here is that the many tens of millions of households that have garages and natural gas hookups will discover the convenience and cost-effectiveness of filling up their clean-burning natural gas vehicles at home.

Postscript: Missed it. Natural gas vehicles gained practically zero market share. Meanwhile electric vehicle adoption is growing exponentially.

The irony is that our natural gas reserves are only abundant if prices are higher. At low prices, extracting shale gas is uneconomical, but at high prices, we have more than we can handle. And such is the trick of the energy trade: having prices high enough to pull the resource out of the ground, but low enough that people can afford to buy it.

Carbon Legislation Will Save the Economy

Five years from now, we'll look back at all the worrying that carbon legislation would ruin the economy and laugh. We worried that U.S. auto companies would be devastated by fuel economy regulations in the 1970s, only to see their resurgence in the years shortly after the rules were put in place. We fretted that cap-and-trade in the 1990s to mitigate acid rain would ruin the economy, only to see the economy grow in the decade after its implementation at the same time sulfur emissions dropped rapidly. The benefits of cap-and-trade outweighed the costs by more than an order of magnitude. And such is the way it will go with carbon legislation.

It's the fundamental American story. There are thousands of hungry entrepreneurs who are just waiting for the right carbon price signal, at which time they will unleash their inventions and services to bring low-carbon energy to our lives. In the process, they will create jobs, wealth and opportunity. And if you include the subsequent ecological and public health benefits of avoided pollution, the economic value will be even greater. The 1930s saw massive global economic depression – and the minting of many millionaires because of great shifts in the energy industry. This era of economic depression will be no different.

Postscript: This one makes me laugh. I really thought we would have carbon policy by 2014. Oh well.

Texas Will Be a Prime Beneficiary of a Carbon-Constrained Economy

One of the great ironies is that Texas, the primary oil state for the last century, will profit handsomely from carbon legislation. In an era when the world will increase its demand for low-carbon fuels, Texas' abundance of solar, wind and natural gas resources will be a jackpot. Its efficient refineries and independent electric grid provide more opportunities to implement sophisticated market mechanisms and sell low-carbon products than anywhere else in the United States. And its geology gives great opportunity to sequester the world's carbon – for a profitable tipping fee, no doubt.

Postscript: Nailed it. Wind, solar, hydrogen, batteries and carbon sequestration are booming in Texas.

We Will See a Serious Water War Somewhere on the Planet

In the next few years, many will die and millions will be displaced because of water scarcity and conflict. Africa, India, the Middle East and interior Asia all are likely locations for this dreaded inevitability. We might as well start preparing now by creating ship-based emergency desalination systems and waste- or solar-powered water treatment systems that our military can deploy as an olive branch and a boost to our foreign policy.

Postscript: I was wrong, thankfully. However, water infrastructure such as dams were targets of ISIS and other violent groups.

2010 Will Be the Year of the Sun

In a more concrete prediction for the next year, I'll argue that 2010 is the year that solar power starts an exponential growth cycle just like wind power did in 2000 – one decade later, but with the same trajectory. The main reason solar will ramp up next year is because it will reach grid parity (it will be cost-competitive with other new power plants) in the near future. That's actually not even really a prediction, as solar is already at grid parity in some places during peak times.

Postscript: Nailed it. Solar is indeed growing exponentially.

A New Mindset

The biggest breakthrough will be that in the next five years, we will once and for all break the intellectual logjam where we pretend we have to choose between the economy and the environment. Instead, we'll choose both.

Postscript: Sadly, no.

Three Cheers for Peak Oil!

EARTH MAGAZINE, JUNE 2009

> *One pithy maxim ("the Stone Age ended before we ran out of stone") captures this sentiment by reminding us that rocks were replaced with something better: metals. And the same story was true with whale oil.*

AFTER DECADES OF BACK AND FORTH, THE DEBATE ABOUT PEAK oil (the year that global oil production peaks then starts an interminable decline) boils down to two points of contention: Is peak oil real, and is it cause for concern? But instead of arguing tired positions that don't seem to be converging on consensus, maybe it's time we shift our tack and instead see what we can do to bring about the peak as soon as possible.

I'm not here to argue whether peak oil is a real problem or not. There are many reasons to believe it is – such as the many prominent geologists talking about a decline in production and the recent run-up in oil prices – and many reasons to believe it's not. After all, the "peakers" have been wrong about the end of oil for more than 100 years, so it's easy to just assume they're wrong again. And we know our total unconventional resource base is massive.

I'm also not going to argue whether we should care if peak oil is near. Sure, essentially everything we value in modern civilization – food, transportation, consumer goods, medicine – will become more limited and expensive if oil production declines and no substitute emerges to take its place. In some extreme scenarios, observers posit that law and order are at risk, governments will have to be redesigned (possibly toward autocracy or other undesirable forms), and economic philosophy will have to be reconfigured.

But our markets might save the day, or something better might come along. One pithy maxim ("the Stone Age ended before we ran out of stone") captures this sentiment by reminding us that rocks were replaced with something better: metals. And the same story was true with whale oil: Despite concerns in the late 1800s that whaling would cause the extinction of these marvelous marine mammals ("peak whale," if you will), we ran out of whale oil customers before we ran out of whales because a competitive product – namely, petroleum-derived kerosene – came along that was better than whale oil for illumination.

Instead of debating this issue, let's do what we can to bring about a peak in oil production and get it over with. Let me explain. About a year and a half ago, I was driving my 8-year-old daughter to her soccer game, about 25 miles away from home. We were stuck in traffic with a seemingly endless stream of cars, when she said, "If everyone keeps driving cars so much, then the world will run out of gas, and that will be great because then we will have to ride bikes, which will be fun."

She's right. Biking would be a lot more fun than sitting in traffic trapped inside a metal cage isolated from the people and nature around us. But it's so easy to get stuck in our patterns and assume that's the way it needs to be without stepping back to question whether that's the way we want it.

Let's be honest – although oil has been an amazing enabler for many things we love (mobility, comfort, healthcare), it's also bad for us in many ways. The rise of our oil-powered modern society has been concurrent with a whole host of problems: divided neighborhoods (split down the middle by highways); asthma epidemics; rampant obesity; petrodollar-funded terrorism; a changing climate. Peak oil might just be the perfect antidote to these social ills.

So, although peak oil sounds intimidating and disastrous, if its arrival is concurrent with a change in our consumptive culture, new behaviors, and the development of an abundant, domestic, renewable, clean, low-carbon alternative, then it can also be synonymous with many good things. Peak oil might mean peak smog. Peak water pollution. Peak obesity. Peak traffic congestion. Peak carbon. We might return to nature hikes, walking or cycling (to my daughter's delight!). Maybe peak oil will be the downfall of drive-thrus and their window-framed relationships, forcing us instead to engage with our fellow citizens face-to face. Can we hope that peak oil will end the construction of highways that divide our neighborhoods by wealth, color or luck? Can we be bold and say peak oil means peak divorce? After all, instead of ferrying ourselves around alone in our cars, maybe we will be home with our families, dining with our spouses, speaking with our neighbors, playing with our children.

I see a post-peak oil world as a place where instead of racing to pump oil out of the ground as fast as possible, we switch gears and decide we're past the petroleum age, and we leave oil – and its dollar value – in the ground as a reserve for a rainy day. If we find a suitable alternative to oil and quit worrying about when its production will decline (thus putting the "peak" behind us), it is possible to imagine a world where, instead of fighting over scraps of oil left below territories controlled by brutal dictators, we make oil irrelevant, and leave these autocrats sitting above boundless reserves of worthless oil, wondering what they did wrong to squander their mineral resource wealth. In these ways, peak oil can help us reduce our foreign entanglements and bolster our diplomacy. Such an approach is in great contrast with today, where our energy trade undermines our foreign policy.

In the end, even though peak oil seems so scary, maybe we should do everything we can to bring it on sooner because that means we will have found something better. And then we can also ride bikes, which will be fun.

> *Postscript: Global oil production *might* have peaked around 2019, though we can't say for certain until we are in the future looking back. It does seem that oil consumption in the USA has peaked and is in steady decline because of a combination of fuel economy standards for automobiles, biofuels mandates, electric cars, and more flexible work-from-home policies in the wake of the COVID-19 pandemic.*

The Bright Future for Natural Gas in the United States

EARTH MAGAZINE, DECEMBER 2012

> *These days, the price of natural gas makes it very attractive to all sorts of end users.*

IN THE LAST YEAR OR MORE, YOU'VE PROBABLY HEARD QUITE A bit about "fracking" – hydraulic fracturing associated with oil and natural gas production from shale formations. Whether you think it's a great new technology, or it's going to ruin the environment, or your opinion falls somewhere in between, the reality is that it and other technological developments have likely forever changed the energy landscape. Thanks to these new developments, we can now affordably produce natural gas from rock formations that previously were inaccessible, and we have more natural gas than ever before. The glut has decreased prices for at least a little while. If recent trends continue – namely, if those prices stay low and various political, environmental and economic pressures to transition to a cleaner, domestic source of energy remain in place, it's likely that over the next decade or two, natural gas will overtake petroleum to become the most popular primary energy source in the U.S.

Postscript: Nailed it.

For a century, natural gas has played an important role in our energy economy. Natural gas has long been valuable for cooking and heating. Over the last few decades, it has also started to play a key role in power generation. But, until recently, it played third fiddle behind coal and nuclear power. Now, that's changing. A few years ago, the amount of electricity generated each month from natural gas surpassed that from nuclear power, and in April 2012, it surpassed coal for the first time in history (even if only for a month). Coal's contribution to electricity generation has dropped from 53 percent in 2003 to about 46 percent in 2011.[59] At the same time, there was a slight drop in overall electricity generation due to the economic recession, so the rise of natural gas came at the expense of, rather than in addition to, coal.

At the same time, oil and petroleum products such as gasoline, diesel and jet fuel practically monopolize the transportation sector, providing about 95 percent of all its energy. Now, natural gas is also eating into that sector too, albeit at a slower pace. A growing number of cities are starting to use bus fleets and trucks that are powered by natural gas, and even some passenger cars can operate on it.

Postscript: Though natural gas buses, garbage trucks and other fleet vehicles were gaining market share, by 2023, new purchases heavily prioritized electric drivetrains because of their lower costs and cleaner operation.

The growing consumption of natural gas is driven by several factors. First, natural gas is considered more environmentally friendly than other fossil fuel energy sources. It has lower emissions of pollutants and

59 The decline of coal and the rise of natural gas has continued since this article was written. Natural gas is now the dominant primary energy source for the power sector.

greenhouse gases per unit of energy than coal and petroleum, and it's less water intensive to use than coal, nuclear energy and biofuels. Second, it has flexible uses across many sectors, as seen in the number of other energy sources it can displace. Third, the U.S. has a lot of it. Domestic production is supplying almost all the annual U.S. demand, and in fact, we have enough of it that if the infrastructure were already built, we could export it.[60] The prospect of making money from natural gas exports is an appealing thought to American energy consumers who send hundreds of billions of dollars overseas every year for petroleum imports. Finally, these days, the price of natural gas makes it very attractive to all sorts of end users.

In contrast, the trends for petroleum and coal are moving the other way. Petroleum use is expected to drop further because of price pressures and policy mandates, such as biofuels production targets and fuel economy standards. Domestic coal use is also likely to drop further because of growing demand from the developing world that will keep prices higher and because of tightening emissions standards expected to rein in coal-borne emissions of heavy metals, sulfur oxides, nitrogen oxides, particulate matter and carbon dioxide.

The biofuels mandates essentially dictate a 0.9 percent annual decline in petroleum use between now and 2022. And ongoing trends imply 0.9 percent annual growth in natural gas consumption for the industrial and power sectors. At those rates, natural gas would surpass petroleum around 2030. If the rates were to double, with a 1.8 percent annual decline in petroleum matched by a 1.8 percent annual increase in natural gas consumption, then natural gas would surpass petroleum in less than a decade.

This scenario is bullish for natural gas, but not implausible, especially for the power sector, in which power plants face stricter air-quality standards and therefore might be put up for early retirement. Coupling

60 LNG exports from the United States began in February 2016: https://www.eia.gov/naturalgas/weekly/archivenew_ngwu/2016/02_25/

all these projections with a reduction in per capita energy use of 10 percent by 2022 along with continued population growth implies that total national energy use would stay the same – a phenomenon that has been true for a decade.

Postscript: Nailed it. It looks like natural gas consumption will surpass petroleum in 2023, a decade after the essay was written. And national energy consumption in the United States today has been level for almost 20 years despite significant population growth and economic growth: This phenomenon indeed turned out to be true.

These positive trends for natural gas do not mean that its use is problem-free. Environmental challenges persist. For example, land surface disturbances from production activity, possible water contamination issues, and induced seismicity from wastewater reinjection are concerns. Emissions from heavy equipment, gas leaks onsite, gas leaks through the distribution system, and flaring at production sites are also issues. However, if those environmental risks are managed properly, then continued growth of natural gas is entirely possible, and natural gas will take the reins from petroleum as America's most popular fuel the same way oil and coal once took over from wood and whale oil.

Stable vs. Volatile Prices

One of the historical criticisms of natural gas has been the relative volatility of its prices, especially compared with coal and nuclear fuels. This volatility is a consequence of several factors, such as large seasonal swings in gas consumption along with the association of gas production with oil, which is also volatile. Thus, large swings in demand and supply can occur simultaneously, but in opposing directions.

Two forces are currently mitigating this volatility, however. First, the increased use of natural gas in the power sector – and not just for direct use in homes for heating – is helping to mitigate some of the seasonal swings as the consumption of gas for heating in the winter might be better matched with consumption in the summer by power plants to meet air conditioning load requirements. And second, because natural gas prices are decoupling from petroleum prices, one layer of volatility is reduced.

Many gas plays are now produced independently of oil production. Consequently, there is a possibility for long-term supply contracts at fixed prices. The combination of these two factors raises the prospects for improved price stability. At the same time, coal, which has historically enjoyed very stable prices, is starting to see higher volatility because its costs are coupled with the price of diesel for transportation. Thus, oil as a driver for price volatility is decreasing for natural gas and increasing for coal.

Postscript: I was right and wrong. Natural gas prices were relatively stable for several years, but with COVID, the February 2021 winter storm in Texas, and the Russian invasion of Ukraine, prices have collapsed and spiked multiple times.

The Complicated Relationship Between Natural Gas and Renewables

The relationship between the use of natural gas and renewables in the power sector is complicated. Renewables are seen as a threat to natural gas, and vice versa, because they compete to make money in the power sector.

Natural gas supporters complain about this rivalry, arguing that policies supporting wind give it an unfair advantage in the competition. Renewable energy supporters counter that gas interests are not required to pay for the pollution they produce – in effect, a form of indirect subsidy itself – and have enjoyed government largesse in one form or another for many decades.

Despite the perception that wind and natural gas are vicious competitors in a zero-sum game, where the success of one must come at the demise of the other, the relationship is more nuanced. In fact, wind and gas benefit one another because they mitigate each other's biggest challenges, thus making each more valuable. The price stability of wind solves the volatility of natural gas. The performance reliability of natural gas solves the variability of wind.

Furthermore, many people seeking a long-term sustainable energy option will often reject natural gas automatically because it is considered a fossil fuel that has a finite resource base. Although most reserves of natural gas were formed many millions of years ago, it is important to note that natural gas can be produced through renewable means. Known as biogas or biomethane, this gas is mostly methane with a balance of carbon dioxide and is created from the anaerobic decomposition of organic matter. Although renewable natural gas is a small fraction of the overall gas supply, it is not negligible.

For example, landfill gas is already an important contributor to many local fuel supplies. And recent studies have noted that the total potential supply available from wastewater treatment plants and anaerobic digestion of livestock waste is over 1 quadrillion Btu annually in the United States.

Postscript: Biogas in fact has grown very slowly as fossil gas has been relatively cheap and easy to produce.

Overall, natural gas has an important opportunity to claim a larger share of the energy market from other primary fuels, displacing coal in the power sector and petroleum in the transportation, industrial, commercial and residential sectors. With sustained growth in demand for natural gas and decreases in demand for coal and petroleum because of environmental and security concerns, natural gas could overtake petroleum as the most widely used fuel in the United States within the next few decades.

Postscript: Nailed it.

Why This Oil Price Collapse Could Be Different

FORTUNE INSIDER, APRIL 2015

> *Unlike the 1980s, electricity prices should hold firm, but we should not be complacent about fixing our energy policy.*

THE [2015] OIL PRICE COLLAPSE SEEMS LIKE A REPLAY OF A BAD 1980s movie that we've seen before. If we are not careful, we'll be doomed to make the same mistakes we made last time, allowing our domestic oil and gas producers to wither, watching energy imports soar, prematurely stunting the growth of alternative fuel sources, and tossing conservation and efficiency by the wayside. If we are smart, we'll seize this opportunity to double down on good energy policies and support all of our domestic energy producers so that we're prepared for the next time oil prices spike.

After the energy crises of the 1970s, there was a brief resurgence of domestic energy production through the mid-1980s. Production grew, wages increased, profits soared, and the Rolls Royce dealership in Midland, Texas, enjoyed a brisk business. But eventually, with OPEC in disarray, Saudi Arabia made the decision to keep its production high to reclaim lost market share, causing global oil prices to plummet in the span of just a few weeks. Sound familiar?

The scars from the 1970s and 1980s run deep. Competing in a vibrant global market, domestic oil companies were gutted by cheaper producers around the world. Our production shrank while consumption increased. Imported oil – much of it from countries whose foreign policy goals do not align with ours – filled in the gap. Our domestic oil and gas operators were in the doldrums for over a decade, with a shrinking, aging workforce and declining production.

Oil producers weren't the only victims. Renewable energy technologies and a cultural mindset of conservation and efficiency got lost too. President Reagan famously removed the solar panels from the roof of the White House, speed limits were increased nationwide, fuel economy standards stagnated, road taxes froze, SUVs grew in popularity, and attitudes about saving energy went the way of bell bottoms. Experimental wind farms and solar panel development slowed down and national R&D funding for energy dropped from a high of $8 billion annually to less than $2 billion.

There are a few reasons why this oil price collapse could be different, but the risks are high that we'll repeat our prior follies.

In the 1970s and 1980s, oil and natural gas prices were closely correlated: When oil prices went up, so did natural gas prices. And, oil was a critical part of the power sector. At its peak in the 1970s, oil fueled one-sixth of all the electrical generation. When oil prices fell, so did electricity prices. Early-stage renewables were too expensive and their policy supports too weak to compete against falling prices, and the renewable power industry came to a screeching halt.

Today the conditions are different. Unlike the 1980s, this time around, electricity prices should hold firm despite low oil prices. In parallel, renewable energy prices have dropped dramatically, making them much more competitive. Those low prices combined with stronger policy supports mean despite what the price collapse did to renewables in the 1980s, renewables are poised to be just fine this time around.

Postscript: Nailed it.

But we still should not be complacent. Low oil prices present us an opportunity to prepare our energy policy for the long term.

One of the greatest outcomes of the shale revolution has been the decline in oil imports. It would be a travesty to give up those gains, returning to a place where we depend so heavily on foreign countries for our energy supply. It will be tempting to loosen up on our energy standards for cars, appliances and light bulbs, but we should continue to prioritize them so that consumers are protected from the price spikes that are surely just around the corner. *(Postscript: Nailed it.)* And we can use efficiency to reduce our consumption alongside any production drops that will occur, helping us keep imports at bay.

Federal R&D investments should be prioritized to help domestic energy producers cut their costs by improving operational efficiencies, minimizing the energy needs for water management at the drilling pad, and capturing fugitive emissions. These investments achieve economic and environmental benefits while helping us keep our domestic producers competitive in a global market.

Now is also the time to put a price on carbon so that we can use the power of markets to combat climate change. Now is the time to fix our broken Highway Trust Fund, reforming an antiquated "gas tax" to include high-mileage, natural gas and electric vehicles, charging vehicles based on the actual damage they cause to the roads.[61]

Policies that encourage the use of clean energy options should be retained, as the low prices for oil might spawn a surge in consumption and emissions, reversing years of gains that have been achieved in national energy efficiency and carbon reductions.

We have a chance to learn from the past. We can watch plummeting oil prices set us back decades, or we can be proactive and use this golden opportunity to implement a long-term energy vision that keeps domestic production high and consumption low while meeting national security and environmental objectives such as reduced imports and emissions. The time for sensible energy policy is now.

Postscript: We did great. Consumption has stayed level despite low prices, growing population and expanding economy, and production has soared. Congratulations to us.

61 See "How to Overhaul the Gas Tax" in Chapter 6, page 156

Break With the Past

MECHANICAL ENGINEERING, JUNE 2020

> *Negative prices and cratering demand:*
> *This oil price crash is different. Now what?*

BOOMS AND BUSTS ARE JUST PART OF LIFE FOR OIL, AS THEY ARE for any other important commodity. Prices go up and they go down. It's a simultaneously thrilling yet ho-hum part of the yo-yo life in the oil patch. In that long view, the latest price crash is just one more paragraph in the long story about black gold. Despite how common price spikes and collapses are, the one this year is different. It has different causes and will have different impacts.

Crude oil prices crashed because demand is decreasing quickly from the pandemic, while at the same time supply increased because of the oil price war between Saudi Arabia and Russia. One or the other is enough to send the markets into a tailspin, but together they are a wicked combination for oil producers in other countries. Those actions, overlaid on limited storage capacity worldwide, caused the price for West Texas Intermediate, one of the world's benchmark crudes, to go dramatically negative for the first time in history. Economic models couldn't handle this situation: Instead of forking over $25 to 75 for the right to own a barrel of crude, producers would pay you handsomely – up to $37 – to take a barrel away.

These are unusual times.

Normally a price drop triggers demand for oil. But this time the demand is dropping for other structural reasons that won't end quickly: not only the COVID-19 pandemic and the rise of video conferencing capabilities as an alternative to travel, but also new policies that prohibit diesel vehicles and promote electric ones, leading to investment by major manufacturers such as Volkswagen into the development of electric drivetrains.

New car designs are executed over multiyear product development cycles, so oil price volatility doesn't really change decision-making for the long run. That means oil demand might not snap right back to where it was before.

> *Postscript: Nailed it. As people returned to work, they didn't all go back to the office daily and oil consumption has not reached prepandemic levels.*

Amazingly, President Trump took the unprecedented maneuver to intervene with the Russians and Saudis to cut oil production and raise oil prices. Think about that: a U.S. president actively sought to use foreign policy as a lever to raise energy prices on American consumers. Why would he do such a thing? While cheap energy is an accelerant for the economy, as a major employer and driver of capital investments, energy production is also important to the economy. Pushing for higher prices protects producers in West Texas at the expense of consumers, but those consumers weren't driving anyway.

Another break from tradition is how oil prices will affect the fate of renewables. According to conventional wisdom, low fossil fuel prices are bad for renewables. The thinking is that higher-priced renewable energy would not be able to compete with the lower-priced fossil options as consumers would choose the more affordable option.

That thinking is getting turned on its head: The oil price collapse might make renewable energy even more attractive. It does from the investor side, not the consumer side.

The renewable energy supply chain has matured a lot in the last few decades, so it's more competitive than ever. Furthermore, while staying at home might cause the demand for transportation fuels to plummet, electricity demand has only dropped a little bit as we consume electrons for our telework, homeschooling and streaming on devices at home.

Perhaps most importantly, low oil prices mean investments in oil and gas are less likely to generate an attractive return. *(Postscript: Was totally wrong. Oil and gas investors have done very well since 2020.)* By contrast, renewables look like a haven investment with fixed prices and guaranteed returns that last two to three decades while sparing exposure to price volatility. Whether consumers want renewables or not, that's what investors want to build, so that's what we'll get.

What do we do now? It is time to put a tax on carbon – price-adjustable to minimize the downward effects of volatility. It's time to invest in efficiency, to reduce consumers' exposure to volatility. And time to invest in R&D, so that the oil and gas sector remains competitive – our oil is the cleanest and most environmentally friendly to produce, so we want to keep it going, but to serve export markets. Last, we should support the transition of workers from oil and gas to other industries, including advanced geothermal energy and carbon management. This price crash is an opportunity: Let's seize it.

Postscript: Did we seize it? Not really. Sigh.

The Coal Industry Isn't Coming Back

THE NEW YORK TIMES, NOVEMBER 2016

DONALD J. TRUMP MADE MANY IMPORTANT CAMPAIGN PROMISES on his way to victory. But saving coal is one promise he won't be able to keep.

Many in Appalachia and other coal mining regions believe that President Obama's supposed war on coal caused a steep decline in the industry's fortunes. But coal's struggles to compete are caused by cheap natural gas, cheap renewables, air-quality regulations that got their start in the George W. Bush administration, and weaker-than-expected demand for coal in Asia.

Nationwide, coal employment peaked in the 1920s. The more recent decline in Appalachian coal employment started in the 1980s during the Reagan administration because of the role that automation and mechanization played in replacing miners with machines, especially in mountaintop removal mining. Job losses in Appalachia were compounded by deregulation of the railroads. Freight prices for trains dropped as a result, which meant that Western coal – which is much cleaner and cheaper than Eastern coal – could be sold to markets far away, cutting into the market share of Appalachian mines. These market forces recently drove six publicly traded coal producers into bankruptcy in the span of a year.

Mr. Trump cannot reverse these trends.

For Mr. Trump to improve coal's fate would require enormous market intervention like direct mandates to consume coal or significant tax breaks to coal's benefit. These are the exact types of interventions that conflict with decades of Republican orthodoxy supporting competitive markets. Another approach, which appears to be gaining popularity, is to open up more federal lands and waters to oil, gas and coal production.

Doing so would only exacerbate coal's challenges, as it would add to the oversupply of energy, lowering the price of coal, which makes it even harder for coal companies to stay profitable. Those same policy actions would also lead to more gas production, depressing natural gas prices further, which would outcompete coal. Instead of being a virtuous cycle for coal, it looks more like a death spiral. And this is all without environmental regulations related to reducing carbon dioxide emissions, which aren't even scheduled to kick in for several years.

Even if the president-elect tried to make these moves, surprising opponents might step in his way. Natural gas companies are the primary beneficiaries of, and now defenders of, clean air and low-carbon regulations. They include Exxon Mobil, the world's largest publicly traded international oil and gas company, which operates in a lot of countries that care about reducing carbon emissions. The company issued a public statement in support of the Paris climate agreement on Nov. 4, the day it took effect. Shutting down coal in favor of natural gas, which is cleaner and emits much less carbon, is a big business opportunity for companies like Exxon Mobil.

In the battle between coal companies and major oil and gas producers, I expect the latter will be victorious.

The rapid uptake of cheap renewables is also a contributor to coal's demise. Mr. Trump made campaign comments suggesting the end of support for renewable energy technologies. But his recent statements call for supporting all energy forms, including renewables, suggesting he won't target them after all.

Even if he did, what are his options? Their tax subsidies are already scheduled to expire or shrink. Plus, wind and solar farms are usually installed in rural Republican districts, which explains why they get so much Republican support in the first place. All those rural districts in America's wind corridor might not be thrilled if their preferred candidate seeks to undermine one of their most important sources of economic growth.

The saving grace for coal production in the United States may be exports to Europe or China. But Europe's demand for coal is waning. And Mr. Trump seems to be marching us toward a trade war with China. Doing so means the Chinese could retaliate by not buying our coal. And even if a trade war is avoided, cheap coal is readily available from nearby Australia.

What does this mean for the average American? More of the same when it comes to energy, which is a good thing. Energy prices will stay low and our air quality will keep improving. And both will help the economy grow.

Any way you slice it, coal's struggles are real and hard to mitigate. No matter how much Mr. Trump tries to protect coal from market competition, doing so will be hard to execute and will get him crosswise with important Republican stakeholders and long-held Republican policy priorities.

Postscript: Nailed it. Coal's decline has continued, dropping from around 730 million tons in 2016 to less than 500 million tons in 2020. Air quality has also improved. And for most of that span, energy prices stayed low (though they increased in 2022 after the Russian invasion of Ukraine).

Michael E. Webber

Will Drought Cause the Next Blackout?

THE NEW YORK TIMES, JULY 2012

> *Trends suggest that water vulnerability will become more important with time.*

THE AMERICAN WEST IS ENDURING A 20-YEAR-LONG megadrought, threatening farmers, their crops and livestock. But there is another risk as water becomes scarcer. Power plants may be forced to shut down, and oil and gas production may be threatened.

Our energy system depends on water. About half of the nation's water withdrawals every day are just for cooling power plants. In addition, the oil and gas industries use tens of millions of gallons a day, injecting water into aging oil fields to improve production, and to free natural gas in shale formations through hydraulic fracturing. Those numbers are not large from a national perspective, but they can be significant locally.

All told, we withdraw more water for the energy sector than for agriculture. Unfortunately, this relationship means that water problems become energy problems that are serious enough to warrant high-level attention.

During the 2008 drought in the Southeast, power plants were within days or weeks of shutting down because of limited water supplies. In Texas, some cities have forbidden the use of municipal water for hydraulic fracturing. The multiyear drought in the West has lowered the snowpack and water levels behind dams, reducing their power output. The U.S. Energy Information Administration recently issued an alert that the drought was likely to exacerbate challenges to California's electric power market, with higher risks of reliability problems and scarcity-driven price increases. And in the Midwest, power plants are competing for water that farmers want for their devastated corn crops.

Unfortunately, trends suggest that this water vulnerability will become more important with time.

Population growth will mean more than 100 million more people in the United States over the next four decades who will need energy and water to survive and prosper. Economic growth compounds that trend, as per capita energy and water consumption tend to increase with affluence. Climate change models also suggest that droughts and heat waves may be more frequent and severe.

Thankfully, there are some solutions.

The government can collect, maintain and make available accurate, updated and comprehensive water data, possibly through the U.S. Geological Survey and the EIA. The EIA maintains an extensive database of accurate, up-to-date and comprehensive information on energy production, consumption, trade and price. Unfortunately, there is no equivalent set of data for water. Consequently, industry, investors, analysts, policymakers and planners lack the information they need to make informed decisions about power plant siting or cooling technologies.

The government should also invest in water-related research and development (spending has been pitifully low for decades) to seek better air-cooling systems for power plants, waterless techniques for hydraulic fracturing, and biofuels that do not require freshwater irrigation.

We should encourage the use of reclaimed water for irrigation, industry and the cooling of equipment at industrial operations like smelters and petrochemical complexes. These steps typically spare a significant amount of energy and cost. The use of dry and hybrid wet-dry cooling towers that require less water should be encouraged at power plants, since not all of them need wet cooling all the time. As power plants upgrade their cooling methods to ones that are less water intensive, these operations can save significant volumes of water.

Most important, conservation should be encouraged, since water conservation results in energy conservation, and vice versa.

New carbon emissions standards can also help save water. A plan proposed by the Obama administration (requiring new power plants to emit no more than 1,000 pounds of carbon dioxide per megawatt-hour generated) would encourage utilities to choose less carbon- and water-intensive fuels. Conventional coal plants, which are very thirsty, exceed the standards proposed by the president. But relatively clean, and water-lean, power plants that use wind, solar panels and natural gas combined-cycle, would meet them. Thus, by enforcing CO_2 limits, a lot of water use can be avoided.

Because rivers and aquifers can span many states (or countries), because there is no alternative to water, and because water represents a critical vulnerability for our energy system, governments at all levels have a stake in working with industry to find solutions. The downsides of doing nothing – more blackouts – are too serious to ignore.

Postscript: Nailed it. One week later, the largest blackout in history, plunging more than 600 million people into the dark in India, was triggered partly by drought. Sadly, conditions have not improved much since this op-ed was written.

Millennials Will Save the World

MECHANICAL ENGINEERING, JULY 2019

Their values-driven, can-do spirit is just what we need.

WHEN I WAS A COLLEGE FRESHMAN, I HAD THE REMARKABLE opportunity to take an introductory level social studies class taught by famed science fiction writer Chad Oliver. He would regale us with stories of the human condition that spanned centuries and continents. One lesson that hangs with me today is that all societies – no matter the time period, wealth, ethnicity or location – have a few things in common. They all have rituals for the dead, such as burials or cremation. They all have other-worldly belief systems, such as religion, witchcraft or superstitions.

And they all complain about the younger generation being lazy and insufficiently respectful of their elders.

It is from this perspective that I want to break with age-old human tradition to thank the Millennial and post-Millennial generations. As a professor who has taught thousands of teens and twenty-something students from around the world, I can tell you that they differ from my generation in one important way: They are better.

Back when I had just received my bachelor's degree, my engineering school peers and I would swap info on our various job offers. Starting pay for good engineering students in the early 1990s was in the low $40,000s per year. Students presented with one offer that paid a salary of $40,000 and another that paid $42,000 would take the latter, no matter who the company was or what it did. All that mattered was the money.

Students today are different. In the United States, Millennials are the first generation who on average will earn less money than their parents. But the ones I know aren't motivated by wealth the same way my college classmates were. Today's new engineers are future-oriented and concerned about sustainability and want to work for a company whose values align with their own. One of my students a few years ago turned down a starting job offer of more than $80,000 per year from a major oil and gas company to take a job that paid roughly $50,000 at a local green utility. In my generation, that was unheard of.

Young people also are rushing to meet our global energy and environmental challenges. In ways both large and small, the Millennial and post-Millennial generations are working to save the world.

Youth often speak with greater moral urgency than older, supposedly wiser generations, who want to appear sensible and take things slowly enough to avoid upsetting the status quo. In the mid-20th century, for instance, student protesters engaged in civil disobedience, sit-ins and freedom rides to animate and empower the Civil Rights movement. High school students who survived the 2018 massacre by gunfire in Parkland, Fla., mobilized a national gun safety movement.

Today, hundreds of thousands of students around the world have been inspired by the example of 16-year-old Swedish schoolgirl Greta Thunberg to stage protests and school walkouts to demand immediate action to stave off the worst effects of climate change.

Time after time, younger generations have taken bold actions after they realize that it's their own future at risk.

My students at the University of Texas Austin will carry the same water bottle around for a year to avoid wasting plastic. They will ride buses or bikes instead of driving. They see climate change much the same way their grandparents saw the Cold War: as a generational challenge and a multidecade battle that must be won to protect our freedoms and prosperity.

It may be fashionable among certain (older) people to bad-mouth the attitude of Millennials and post-Millennials, in contrast to the can-do spirit of those past generations, but in my experience, they are as hard-working and public-spirited as anyone.

Thank goodness. We need them. The energy and environmental challenges today's young people face are every bit as stark as World War II and the Cold War. The causes are deeply embedded in the global economic system, and the effects impact everyone one way or another.

It's naive to think that youthful activism alone will save us from the long-lasting pollution of our past and present choices. But I'll take the Millennials' can-do spirit over a turn-a-blind eye, take-it-slow negligence any day.

Postscript: Too early to judge my prediction on this one, but I'm optimistic that I'm correct.

Climate of Optimism

MECHANICAL ENGINEERING, FEBRUARY/MARCH 2022

> *After a year of droughts and storms and fire and floods, it is easy to despair. But the case for hope is straightforward: We will solve the climate challenge because customers demand it, employees demand it, and investors demand it.*

A FAMOUS 2008 COMIC PANEL BY THE ARTIST RANDALL MUNROE features a stick figure man furious at a computer. (Munroe's characters are all stick figures.) Asked whether he was coming to bed, the character replies that he can't because, "Someone is wrong on the Internet."

In addition to my teaching, research and writing, I post on social media because it's a way to reach an entirely different audience. And with some of the responses I get, it is easy to imagine a person like Munroe's stick figure pounding on a keyboard or jabbing at a phone screen. It shouldn't be a surprise really. Climate change and the energy transition – the subject matter of my typical musings – are topics that evoke a lot of emotion. As odd as it may seem, posts about nuclear energy and heat pumps consistently generate passionate reactions.

Last summer, I posted a simple comment on Twitter that (as the kids say) blew up, garnering more than 200,000 impressions. Someone asked his followers about a personal belief that people inside one's own political circle would find ridiculous. I replied, "I believe with great clarity that we will, 1) solve the climate crisis, 2) faster than people anticipate and 3) remove CO_2 from the atmosphere to get us to preindustrial levels."

My climate optimism – and the opinion that this best-case outcome is already baked in – provoked disbelief and a request for an explanation. Maybe that is understandable: The headlines are full of gloom, from nations failing to meet even their most watered-down emissions goals, to the fires and floods that hint at what a future with an altered climate could mean. To me, though, the case is straightforward. We will solve the climate challenge because customers demand it, employees demand it, and investors demand it. Many (but obviously not all) policymakers at municipal, state, regional and national levels are pushing for it, and those recalcitrant ones in opposition are slowly but surely losing hold.

It's an argument that many people have a hard time accepting. Climate change seems like a challenge perfectly designed for humanity to fail. To fight it requires coordinated action of hundreds of nations and billions of people, all working toward a seemingly abstract goal that will only be reached decades from now. In a nation where some people don't have the patience for instant ramen, that sort of delayed gratification seems alien.

Yet, the world has faced difficult global challenges before and overcome them, from World War II and the Cold War to the thinning of the polar ozone layers. For instance, people tend to forget how intractable the Cold War was. By the 1970s, it seemed less like a resolvable conflict than a permanent state of world affairs. Both sides were unwieldy, multicultural coalitions that invested large chunks of their national incomes on weapons systems and prestige projects for decades on end.

Given the demands, those coalitions could have splintered, but they held in place for more than 40 years until the fall of the Berlin Wall. Similarly, the growing hole in the ozone layer looked as if it were a fact of life in the mid-1980s, but then the nations of the world swiftly agreed to phase out production and use of the chemicals implicated in depleting ozone. There's a long way to go in restoring the ozone layer, but the multinational commitment to stabilizing it and reducing the damage has been phenomenal.

Climate change is a multinational, multidecade problem and we can solve this one too. The signs of progress are found in some surprising places.

Action Is Going to Happen – and Fast

Often, when people express the opinion that climate action will win out over denial or delay, they point to the salience of this issue with younger people. And it is true, at least in my experience, that concerns about climate change are often bifurcated by a generational divide. While older people can afford to be more skeptical and less concerned, people 50 years old and under – the people who will have to live with a warmer future – are more passionate about acting.

Passion by itself isn't enough. In my teaching, I have the great privilege and honor to work with students from around the world. They are in despair by the mess older generations are bequeathing them and are determined to fix it. And they will fix it: Believe me when I tell you they work hard, they're smart, they're focused. Importantly, they are starting to take leadership positions everywhere to have the power to act.

If the only evidence I had was the rising generations' focus on the challenge, I might be less confident than I am. After all, the time to make the necessary changes is not in decades' time, when the current generation of leaders have relinquished power, but now and in the next few years. To me the best indicator that action is going to happen – and fast – is that energy companies themselves are now pushing for change, somewhat quietly now, but evermore noisily as it's clear that this is what the markets are calling for.

As a professor who advises a major clean-tech venture fund and spent three years serving as chief science and technology officer for ENGIE, one of the world's largest energy companies, I get to see what the corporate world is doing. I also teach executive education courses for the energy industry and get to see what early- to midcareer engineers and business leaders want. I know the pressure they all are putting on their employees to reduce emissions. For instance, ENGIE has pledged to get to net-zero by 2045 for not only its own direct emissions, but also all emissions upstream and downstream from its activities. (I bet they hit that target early.) Other companies, under pressure from regulators, investors, employees and customers, will follow suit.

That pressure from investors is real. It is a growing trend for investors to include nonfinancial criteria for their investment decisions. These criteria include factors such as transparency, ethical operations, and fair labor treatment as well as sustainability. Investment determined by these environmental, social and governance (ESG) criteria is anticipated to grow quickly. By one estimate, there will be $50 trillion worth of ESG assets by 2025, which is a third of the assets under management worldwide.

The result? Getting money for a coal mine in the United States is hard, and it is becoming more difficult and more expensive to get money for new large-scale hydrocarbon production.

Oil and gas companies know this, and rather than sit tight and watch as ESG investors abandon them, they are looking to pivot to opportunities that leverage their specialized skill sets for a low-carbon future. Some of these businesses include advanced drilling and subsurface mapping for geothermal energy; carbon management through capture, utilization and sequestration; hydrogen production, transportation and applications; and offshore capabilities for ocean-based wind and solar production.

This isn't conjecture on my part. Last summer, a multibillion-dollar exploration and production company from the oil and gas sector called me to get some insights. In their words, they see five to six years of activity left for them, not two or three decades. For them, speed is critical: They want to make oil and gas better, make money making oil and gas better, then use that money to get out of oil and gas entirely.

Some European oil and gas companies, such as Ørsted, have already made the change. Formerly known as Danish Oil and Natural Gas, the company is now the world's largest developer of offshore wind power. Others such as Equinor (formerly known as Statoil, the state-owned oil company of Norway) are in transition now.

I also get phone calls or emails at least once per week from representatives of huge entities – corporations you recognize, multibillion-dollar companies, some of the most important hydrocarbon families in American history – asking about how to catch up with the energy transition. As one caller put it, "The train left the station." These third- and fourth-generation fossil fuel families don't want to be the ones that lose the family fortune clinging to a legacy. They want to be part of the future and understand that the future is low carbon. They feel tardy and in a hurry.

Significant levels of investment are already flowing, quietly, and even more is about to flow loudly for renewables, storage, carbon capture utilization and sequestration, hydrogen, new infrastructure and more. The amount of money is phenomenal. To my eye, these trends are irreversible. It's an avalanche that has already started and can't be stopped or turned back.

The Only Question Is Who Will Lead

I work with a $3 billion clean-tech venture capital fund, helping them to screen technologies and identify entrepreneurs. From that seat I can't help but feel optimistic: There are a lot of great ideas out there. Wind and solar have fallen in price so much in the last decade or two that in some situations it is less expensive to build new zero-carbon generating assets than to keep existing – and already paid-for – coal plants up and running. Without breaching a nondisclosure agreement, I can tell you other technologies, such as novel ways to produce hydrogen, batteries with innovative chemistries, clever electrified transportation technologies, and large-scale thermal storage are poised to make the same kind of impact. We just need to clarify the market rules, open the spigot of investment, then watch them go.

Pundits who look at the trajectory of carbon emissions over the past several decades and throw up their hands in despair should look at the environmental problems of the 1960s, 1970s and 1980s, things like the persistent smog of Los Angeles, the flammable Cuyahoga River in Cleveland, acid rain falling on the Eastern United States, and the ozone hole. At first, riled-up citizens were at loggerheads with industry leaders, while policymakers sat passively. In time, however, solutions were found, and reclamation could begin. Today, each of those environmental crises is markedly improved.

We will reclaim the atmosphere to its preindustrial greenhouse gas composition too. That means going beyond net-zero emissions. We need to go carbon negative – by taking CO_2 out of the atmosphere

– to clean up 200 years of emissions. Pollution in solid and liquid forms led to the creation of waste management industries and water and wastewater treatment systems in the United States that are worth a few hundred billion dollars annually and employ hundreds of thousands of people. Climate restoration will compel the building of an atmospheric carbon management industry that will generate jobs and grow the economy.

But what does my climate optimism mean for today's fossil fuel companies? It depends on whether they have the capacity to work with, rather than against, the energy transformation.

On the one hand, electricity, renewables, materials, data and mining companies will thrive. Remaking our economy is a multitrillion-dollar opportunity to clean up our act, and companies in these spaces will play a big part.

Conversely, demand for coal will be reduced to critical needs, primarily metallurgical and material purposes such as making steel, carbon filters and cement. A few coal companies can survive on that, but the days of coal serving as the dominant fuel for the world's power grid are over.

Oil and gas can have a great future, I believe, but only if the industry chooses to leverage its expertise for cleaner options.

It is true that there are some American domestic independent oil and gas companies that are holding out against this tide. They will struggle to get financing or to recruit and retain talent. The best young workers out there have options, and from what I have seen, they would rather choose to work for a tech company that pays better, has a net-zero pledge, and whose shares are gaining value rather than working for a company that they perceive to be part of the problem and has declining market capitalization.

The decline in share prices isn't theoretical. Over the past ten years, Standard and Poor's Oil & Gas Exploration & Production Select Industry Index has witnessed an annualized decline of 6.67 percent. That compares to an annualized increase of 6.84 percent for utilities and an annualized increase of 12.26 percent in the S&P 500, suggesting that the woes with oil and gas aren't a problem with the energy industry at large.

Postscript: In 2022, this trend reversed, as market capitalizations for oil and gas companies improved and sagged for tech companies.

Ultimately, companies want to make money, and the oil and gas sector has not been a moneymaker or value-creator in recent history the way it had been for the century before that. As oil and gas companies' stock prices stagnate, employees delay retirement, keeping payrolls top-heavy and making it harder for companies to replenish their workforce with fresh ideas.

Climate solutions are coming fast because customers demand it, investors demand it and employees demand it. The pathway is clear and baked in. Companies that ignore these pressures do so at their peril.

The only question now is which companies – and which countries – will lead.

The United States can still lead. Texas, a state whose identity is tightly bound to the petroleum industry, could still lead. The countries and companies that move the fastest and smartest will make a lot of money. The deniers and laggards will lose. As an American and a Texan, I hope we don't side with the laggards.

But even if Texas and the nation stupidly pass up the climate action opportunity, someone in the world will still solve it. That much is clear. Believe me, we've got this. The swiftness of action to halt and reverse the climate crisis will be stunning. And it gives me hope.

That's why I am a climate optimist.

Postscript: Check back with me in five years to see if this prediction is correct – or if I'm still a climate optimist!

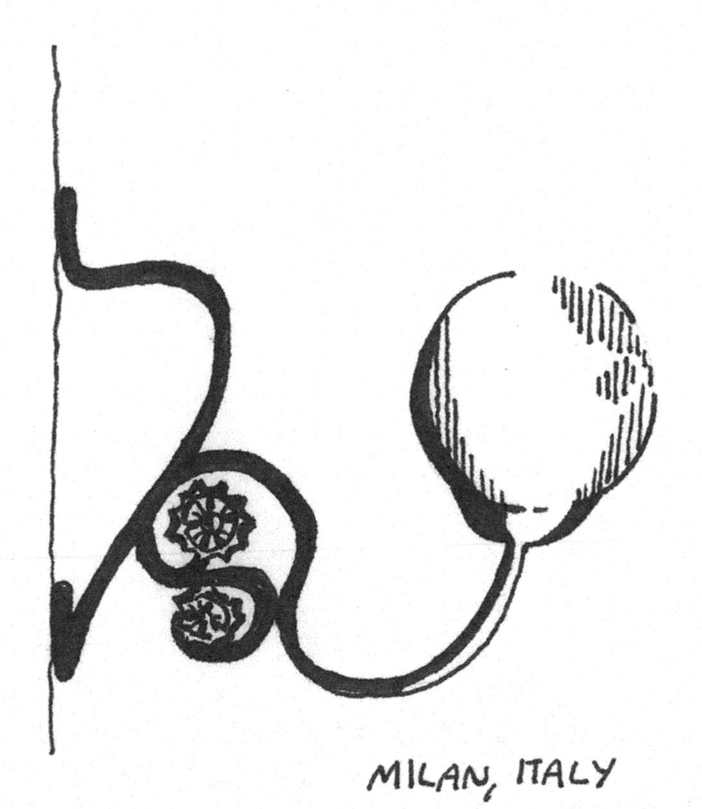
MILAN, ITALY

Bibliography

This collection is based on a subset of essays, articles and op-eds that I authored or co-authored between 2008 and 2022. Each piece and its publication information are listed below.

Prologue

Parts of the prologue were adapted from:

Michael E. Webber, *Thirst for Power: Energy, Water and Human Survival*, Yale University Press (2016).

Michael E. Webber, *Power Trip: The Story of Energy*, Basic Books (2019).

Michael E. Webber, "Invisible Heroes," *Mechanical Engineering*, May 2020.

Chapter 1

Michael E. Webber, "Our Future Rides on Our Ability to Integrate Energy + Water + Food: A puzzle for the planet," *Scientific American*, February 2015.

Kelly T. Sanders and Michael E. Webber, "The Energy-Water Nexus: Managing water in an energy constrained world," *EARTH Magazine*, July 2013.

Michael E. Webber, "What's the World to Do About Water?" *Popular Science*, June 2014.

Michael E. Webber, "Our Water System: What a Waste," *New York Times*, March 22, 2016.

Michael E. Webber, "The Water Trade: In a world of declining freshwater availability, we are exchanging energy for clean water to meet the needs of thirsty billions," *Mechanical Engineering*, November 2016.

Sheril R. Kirshenbaum and Michael E. Webber, "Time for Another Giant Leap for Mankind," *Issues in Science and Technology*, Spring 2012.

Michael E. Webber, "LISTEN: What's the Carbon Footprint of Your Thanksgiving Dinner?" *KUT* Radio, Austin, TX, November 27, 2013. [www.kut.org/energy-environment/2013-11-27/listen-whats-the-carbon-footprint-of-your-thanksgiving-dinner]

Chapter 2

Alex C. Breckel, John R. Fyffe and Michael E. Webber, "Trash-to-Treasure: Turning nonrecycled waste into low-carbon fuel," *EARTH Magazine*, August 2012.

Michael E. Webber, "Tapping the Trash: Transforming costly wastes into valuable resources can make cities highly efficient," *Scientific American*, July 2017.

Michael E. Webber, "Wasting an Opportunity: Americans generate a lot of trash and stick most of it in landfills," *Mechanical Engineering*, November 2014.

Michael E. Webber, "The Carbon Dioxide We Dump Into the Sky Is Just Another Kind of Garbage," *Scientific American*, December 17, 2019.

Joshua D. Rhodes and Michael E. Webber, "The Solution to America's Energy Waste Problem," *Fortune*, December 18, 2017.

Chapter 3

Michael E. Webber, "We Need More Than STEM: The energy sector's challenges require education in the arts as well as science, math, and technology," *Mechanical Engineering*, January 2020.

Michael E. Webber, "Better Tools for Energy Literacy: The old methods of teaching – using chalk and talk – aren't giving citizens a deep enough understanding of complex energy issues," *Mechanical Engineering*, April 2014.

Michael E. Webber and Sheril R. Kirshenbaum, "It's Time to Shine the Spotlight on Energy Education," *The Chronicle of Higher Education*, January 22, 2012.

Michael E. Webber, "New Engineering Thinking for a New Climate," *Mechanical Engineering*, June 2016.

Chapter 4

Michael E. Webber, "This Earth Day, Pandemic Offers Opportunity to Fix Our Air and Water Woes," *Austin American-Statesman*, April 22, 2020.

Michael E. Webber, "Turning Around Our Priorities: We have a few things backward that we should reverse for energy and climate," *Mechanical Engineering*, March 2020.

Michael E. Webber, "The Oil Industry Is Part of the Solution: Activists want to punish the petroleum industry, but it has the technical chops to tackle climate change," *Mechanical Engineering*, February 2020.

Michael E. Webber, "How Cheap Gasoline Hurts the Environment," *Dallas Morning News*, June 8, 2016.

Michael E. Webber, "Forum: Include Agriculture in Emissions Policy," *Corpus Christi Caller-Times*, March 5, 2015.

Chapter 5

Michael E. Webber, "An Invisible Hand Driving Energy Policy: The finance and insurance industry has a major say in how we address mobility and climate issues," *Mechanical Engineering*, April 2020.

Michael E. Webber, "Over the Hills and Far Away: With global supply chains locking up, it's worth looking to the energy industry for some lessons," *Mechanical Engineering*, December 2021/January 2022.

Michael E. Webber, "Boomtown," *The Alcalde*, July/August 2014.

Michael E. Webber, "A Dirty Secret – China's Greatest Import: Carbon Emissions," *EARTH Magazine*, January 2011.

Michael E. Webber, "Conflict Between Russia and Georgia Adds New Twist to the Energy War," *Austin American-Statesman*, August 17, 2008.

Chapter 6

Michael E. Webber, "Fly More, Not Less: There's no shame in flying. But we must find ways to reduce aviation's climate impact," *Mechanical Engineering*, December 2019.

Michael E. Webber, "A New Age of Rail: Reinvesting in freight railroads could be an infrastructure solution to multiple challenges," *Mechanical Engineering*, February 2018.

Michael E. Webber, "Today's Decisions, Tomorrow's Cities: Is building car-centric infrastructure a dead end?" *Mechanical Engineering*, October 2019.

Michael E. Webber, "Electric Highway: A revolution in electrified, autonomous vehicles is poised to change the way we get around – and the very landscape we travel through," *Mechanical Engineering*, June 2019.

Michael E. Webber, "Can We Escape Our Car-Centric World? Freedom has become just another word for nothing left to choose," *Mechanical Engineering*, April 2019.

Michael E. Webber, "This Pandemic Could Cause a Long-Term Shift in Car Ownership," *Milken Institute's Power of Ideas*, July 15, 2020.

Michael E. Webber, "How to Overhaul the Gas Tax," *New York Times*, December 24, 2013.

Chapter 7

Michael E. Webber, "What to Do About Natural Gas: The massive gas infrastructure could pose a barrier to decarbonizing our energy system, but it doesn't have to. Here's how," *Scientific American*, April 2021.

Michael E. Webber, "It's Complicated: It's time we figure out our relationship with nuclear energy," *Mechanical Engineering*, June 2019.

Michael E. Webber, "Renaissance and Revolution: Nuclear power's long-delayed revival is a victim of the rise of shale gas," *Mechanical Engineering*, April 2018.

Michael E. Webber, "Making Renewables Work: Grid operators face real challenges in integrating wind and solar power. Maybe we need to rethink the relationship between electricity supply and demand," *Mechanical Engineering*, December 2016.

Joshua D. Rhodes, Michael E. Webber, Thomas A. Deetjen and F. Todd Davidson, "Are Solar and Wind Really Killing Coal, Nuclear and Grid Reliability?" *The Conversation*, May 11, 2017.

Michael E. Webber, "How Dependable Is the Traditional Grid? Utility managers worry that renewables may hurt grid reliability. In fact, all energy forms and technologies suffer from some sort of reliability challenge," *Mechanical Engineering*, October 2016.

Chapter 8

Michael E. Webber, "Innovation Should Know No Boundaries: American industry – especially the energy industry – is stronger when it can draw talent from all over," *Mechanical Engineering*, January 2019.

Michael E. Webber, "The Case for Cross-Sectoral Disruption: When the innovations come from unexpected directions, we need as many points of view as possible," *Mechanical Engineering*, August/September 2021.

Yael R. Glazer, Jamie J. Lee, F. Todd Davidson and Michael E. Webber, "Shale Boom Could Fuel Batteries," *EARTH Magazine*, April 2017.

Michael E. Webber, "Breaking the Energy Barrier: Can the Department of Defense solve the world's energy crisis one jet at a time?" *EARTH Magazine*, September 2009.

Michael E. Webber, "Research Is Necessary to Accelerate Our Transition to a Zero-Carbon World," *Pour La Science* (the French edition of *Scientific American*), April 2020.

Michael E. Webber, "Build First, Explain Second: We can ramp up on deployment of clean energy systems, and then apply what we learn to make the whole system better," *Mechanical Engineering*, June/July 2021.

Michael E. Webber, "Energy Is All Around Us, Including Up Above and Down Below," *Pour La Science* (the French edition of *Scientific American*), May 2021.

Chapter 9

Michael E. Webber, "Redefining Humanity Through Energy Use," *EARTH Magazine*, March 2010.

Michael E. Webber, "The Color of Energy: The industry must work to solve its racial disparities," *Mechanical Engineering*, August 2020.

Michael E. Webber, "Setting the Direction: For women to take a more equal leadership role in the energy industry, policies need to support families," *Mechanical Engineering*, September 2019.

Michael E. Webber, "The Farm Woman's Dream, 100 Years Later," *The American Society of Mechanical Engineers Energy Blog*, June 29, 2020.

Michael E. Webber, "How Clean Energy Can Win Over Rural Areas: Building out renewable energy infrastructure can bring opportunities to places eager for jobs," *Mechanical Engineering*, February/March 2021.

Michael E. Webber, "Learning How to Beat the Heat: As the world gets hotter, places will need to adapt their infrastructure – and long-held cultural traits," *Mechanical Engineering*, August 2019.

Chapter 10

Michael E. Webber, "Industry Needs a Better Approach to Communication," *Houston Chronicle*, March 24, 2015.

Michael E. Webber and Sheril R. Kirshenbaum, "How `Frankenstein' Prevents Us From Tackling Climate Change," *EARTH Magazine*, March 2016.

Michael E. Webber, "A Pitch to Study BREW: The Beer-Renewable Energy-Water Nexus," *EARTH Magazine*, September 2017.

Michael E. Webber, "World War G: Zombies, energy and the geosciences," *EARTH Magazine*, December 2013.

Michael E. Webber and David Brown, "What You Might Have Missed During 'The Big Game'," *KUT Radio*, Austin, TX, February 11, 2014. [www.kut.org/energy-environment/2014-02-11/what-you-might-have-missed-during-the-big-game]

Michael E. Webber, "Energy and Christmas: Not Just a Lump of Coal in Our Stockings," *Unpublished*, written December 21, 2015.

Chapter 11

Michael E. Webber, "The Risky Business of Predicting the Future: Mistaken forecasts can limit our progress," *Mechanical Engineering*, December 2018.

Michael E. Webber, "A Fool's Look Into the Future," *EARTH Magazine*, December 2009.

Michael E. Webber, "Three Cheers for Peak Oil," *EARTH Magazine*, June 2009.

Michael E. Webber, "The Bright Future for Natural Gas in the United States," *EARTH Magazine*, December 2012.

Michael E. Webber, "Why This Oil Price Collapse Could Be Different," *Fortune*, April 27, 2015.

Michael E. Webber, "Break With the Past: Negative prices and cratering demand: This oil price crash is different. Now what?" *Mechanical Engineering*, June 2020.

Michael E. Webber, "The Coal Industry Isn't Coming Back," *New York Times*, November 16, 2016.

Michael E. Webber, "Will Drought Cause the Next Blackout?" *The New York Times*, July 24, 2012.

Michael E. Webber, "Millennials Will Save the World: Their values-driven, can-do spirit is just what we need," *Mechanical Engineering*, July 2019.

Michael E. Webber, "Climate of Optimism," *Mechanical Engineering*, February/March 2022.